THE ABM TREATY AND
WESTERN SECURITY

This book was produced under the auspices of the Defense and Arms Control Studies Program, Center for International Studies Massachusetts Institute of Technology

THE ABM TREATY AND
WESTERN SECURITY

WILLIAM J. DURCH

BALLINGER PUBLISHING COMPANY
Cambridge, Massachusetts
A Subsidiary of Harper & Row, Publishers, Inc.

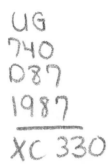
International Standard Book Number: 0-88730-264-5

Library of Congress Catalog Card Number: 87-26978

Printed in the United States of America

Library of Congress Cataloging-in-Publication Data

Durch, William J.
 The ABM Treaty and Western security.

 Bibliography: p.
 Includes index.
 1. Ballistic missile defenses. 2. Soviet Union. Treaties, etc.
United States, 1972 May 26 (ABM). 3. Nuclear arms control.
4. Europe—National security. 5. United States—National
security. I. Title.
UG740.D87 1987   358'.1754   87-26978
ISBN 0-88730-264-5

CONTENTS

About the Author

ACKNOWLEDGMENTS

This work grew out of a request by Robert Nurick and the late Jonathan Alford of the International Institute for Strategic Studies, London, in April 1983 for a study of ballistic missile defense. It developed into a book made possible by research support from the Defense and Arms Control Studies Program at the Center for International Studies, Massachusetts Institute of Technology. I would like to thank the program's director, Jack Ruina, for his unstinting support and encouragement of the project.

A number of people took time from busy schedules to read and comment on the draft manuscript. I would like to extend special thanks to Barry Blechman, Matt Bunn, Ivo Daalder, Lynn Davis, Charles Gellner, Cristann Gibson, Charles Glaser, Sidney Graybeal, Thomas Karas, Glenn Kent, Herbert Lin, John McNeill, Stephen Meyer, John Rhinelander, Condoleeza Rice, and James Schear for their painstaking attention to the devils in the detail and their very helpful advice. For their reviews of an earlier draft I would also like to thank Antonia Chayes, George Rathjens, Russell Shaver, and Charles Whitechurch. Any remaining errors or omissions in the final product are, of course, the responsibility of the author.

Finally, I would like to thank my wife, Jane, for her support and her patience with a project that has steadily consumed evenings and weekends for far longer than either of us care to remember.

INTRODUCTION

The United States is in the midst of its second great debate over the feasibility and desirability of strategic ballistic missile defense (BMD). The first debate terminated in May 1972 with the signing of the SALT I, U.S.-Soviet Anti-Ballistic Missile Treaty (the ABM Treaty), which strictly constrained the development and deployment of strategic BMD.[1] The ABM Treaty has no expiration date but every five years comes up for a formal review that is designed to focus the parties' attention on the agreement, on the other party's performance and on changes that might need to be made. The third five-year review period opens in the fall of 1987.

Since the 1982 review, the Reagan administration has embarked upon the Strategic Defense Initiative (SDI), an effort to determine whether, in President Reagan's words, nuclear ballistic missiles can be rendered "impotent and obsolete." Many advocates of arms control feel that SDI is an effort to render arms control, at the heart of which lies the ABM Treaty, impotent and obsolete as well. The debate triggered by SDI includes issues of technical feasibility, "crisis stability," and the arms race itself, but it goes beyond them to encompass more fundamental issues about the nature of the Soviet Union and the desirability of U.S.-Soviet political and strategic parity. At the core of the debate is an argument about whether Western security irretrievably rests in Soviet hands. It is in some respects an old debate, about fear of the Soviet Union versus fear of nuclear weapons and the proper balance to strike between them in forming U.S. public policy. Implicit in SDI, however, is a promise to eliminate the need for such balance; to protect against both fears by shifting U.S. nuclear strategy from deterrence to defense.

Undersecretary of Defense Fred Iklé, who is a partisan of SDI, once asked whether nuclear deterrence could last out the century.[2] Over fourteen years, half way to that goal, have elapsed since then. Deterrence has, apparently, stood the test of détente, weak leadership on both U.S. and Soviet sides, new generations of weapons, and the new Cold War. It has survived periods of high and low U.S. defense spending, the depths of Watergate and the rise of Solidarity, MIRVing, and the opening and closing of the "window of vulnerability." The potential contribution of strategic BMD to maintaining that resilience is one of its potential attractions. But the relevant questions remain: what contribution to what end; at what cost politically, strategically, and economically; and to the exclusion of what other options? Some fear that one excluded option could be arms control.

Even if there were no SDI, however, the ABM Treaty would still need refurbishing. New technologies with potential applications to ballistic missile defense are coming over the horizon, and incremental improvements in older technologies are giving them near-ABM capabilities, thus threatening to undermine the basis of the Treaty. Soviet compliance practices have raised problems, and basic ambiguities in its text also need to be resolved.

What follows is an effort to sort through the problems and prospects of strategic defense and the ABM Treaty, starting from the current framework and working toward future policy options. Chapter 1 provides a sense of historical perspective by reviewing the history of U.S. and Soviet strategic defense and its relationship to strategic doctrine and offensive forces. Chapter 2 looks at arms control and the Treaty itself, at its costs and benefits, and at its current problems. Chapter 3 examines three different paths for U.S. policy: termination/replacement; modification to permit limited missile defenses; and a strengthened Treaty regime. Chapter 4 compares and contrasts these three paths, discusses policy options for review of the Treaty, and offers conclusions.

NOTES

1. SALT, an acronym for Strategic Arms Limitation Talks, is shorthand for U.S.-Soviet bilateral negotiations on strategic nuclear arms control from November 1969 through June 1979. *SALT I* refers to the negotiations up to May 1972 and, by convention, to the agreements they produced, including the Interim Agreement on offensive strategic ballistic missile forces and the ABM Treaty. *SALT II* refers to follow-on negotiations that culminated in the treaty of June 1979, which was signed by the United States and the Soviet Union but not ratified.

   ABM was the common shorthand for ballistic missile defense when the SALT I agreements were signed. The term can be used interchangeably with *BMD* although in contemporary usage ABM is used almost exclusively in reference to the Treaty and its provisions. Finally, the terms *strategic defense* and *SDI* are not synonymous; the former is a general term that can encompass air defense and civil defense (sheltering) as well as BMD, whereas the latter term refers to a specific U.S. BMD research and development program.

2. Fred C. Iklé, "Can Nuclear Deterrence Last Out the Century?" *Foreign Affairs* (January 1973).

1 STRATEGIC DEFENSE IN HISTORICAL CONTEXT

The contemporary debate over strategic defense and arms control is driven in part by notions, some based in fact, some not, about U.S. and Soviet behavior, nuclear forces, and nuclear doctrine; these cannot be evaluated properly without first gaining an appreciation of their historical contexts. Thus, before examining the ABM Treaty and the future of strategic defense, the past four decades of strategic defense efforts are reviewed briefly in the context of evolving U.S. and Soviet doctrine and actual offensive capabilities.

OFFENSE AND DEFENSE, 1945–1982

The United States emerged from World War II as the sole possessor of the atomic bomb and the world's premier industrial power. The bomb assumed increasing importance in U.S. strategy as the United States realized that it could never again step back from involvement in European affairs, as it had after World War I. Soviet interests and Soviet power made it clear that the United States would remain entangled in European affairs for some time to come. U.S. nuclear weaponry became the principal counter to the Soviet Red Army. For the first decade of the nuclear age, the reverse was also true: The Soviet Union's roughly two dozen divisions in Eastern Europe comprised a conventional military threat to Western Europe, thus countering U.S. strength based on possession of the bomb. After 1954, that threat extended to include Soviet nuclear weapons and, by 1956, nuclear-capable aircraft with intercontinental range. By the late 1950s, U.S. territory was facing a serious military threat for the first time since the War of 1812.

However, U.S. perceptions of threat predated actual Soviet capabilities. U.S. intelligence on those capabilities was poor, so policy was based on worst-case assumptions.[1] Soon after the Soviet Union exploded its first fission bomb in August 1949, the United States began to improve its air defenses as a hedge against one-way missions by Soviet medium-range bombers. In early 1950, well before the outbreak of the Korean War, there was a flurry of state and federal interest in planning for civil defense.[2] Strategic defense did not become a high national priority, however, until the final month of the Truman administration. The Eisenhower administration's Basic National Security Policy (NSC 162/2) of October 1953 called for development of "an integrated and effective continental defense system." The follow-on policy (NSC 5408) called for a continental defense system that would "minimize the effects of any Soviet attack" and prevent "the threat of atomic destruction from discouraging U.S. freedom of action or weakening national morale." Continental defense, then, was intended to serve both military/strategic and foreign policy/psychosocial ends. The policy called not only for air and civil defense, but for measures to prevent the "clandestine introduction and detonation of atomic weapons," "to determine known subversives for detention in the event of emergency," and to assure the security of ports. Looking to the future, it stressed that "changing the metropolitan pattern of America so that it presents fewer concentrated targets for attack may be essential in the age of intercontinental ballistic missiles."[3] Like many a policy-cum-advocacy document, its reach sometimes exceeded its grasp. Nonetheless, it gives evidence of the scope of strategic defense contemplated in the mid-1950s—a program much more comprehensive than that contemplated under the Strategic Defense Initiative (SDI), a program devised to cope not with a standard military threat but with a perception of nuclear-armed malevolence.

In August 1954, the U.S. Air Force was given purview of all continental air defense forces. In the defense reorganization of 1958, the Air Defense Command became a unified (multiservice) command reporting directly to the Joint Chiefs of Staff (JCS). By the end of that decade, new fighter-interceptors, surface-to-air missiles (SAMs), radar fences, and command and control systems had been built and deployed. The Semi-Automatic Ground Environment system (SAGE), the Air Force's primary air defense command and control network, stimulated the development of digital computers to process and coordinate tracking data from the system's far-flung radars. SAGE was managed from three regional combat control centers, which in turn oversaw twenty-three "direction centers" housed in four-story blockhouses, some of which were located on Strategic Air Command (SAC) bases. The first SAGE direction center became operational in 1958; five years later the system was fully operational. However, by 1960 U.S.plans for the fully operational system had been scaled back from 4,500 fighter-interceptors to 1,000, and from nearly 11,000 long-range SAMs to less than 3,000. The large Soviet bomber threat predicted in the mid-1950s had failed to materialize. After building fewer than 200 heavy bombers, the Soviet Union instead had concentrated its strategic nuclear offensive efforts on the development of intercontinental ballistic missiles (ICBMs).[4]

The Soviet Union also made a major effort in the area of bomber defense. In 1948, the Troops of National Air Defense (Voyska Proti-vovozdushnoy Oborony Strany or PVO Strany) became a separate armed service and in 1954 received its own commander in chief. Its sole mission was the air and missile defense of the Soviet homeland. Through the mid-1950s, its fighter-interceptor forces consisted solely of clear-weather jet fighters. Starting in 1955, these were supplemented by SA-1 SAMs placed in fixed sites around Moscow. Beyond range of the SAMs (roughly eighty-five miles from Moscow), U.S. SAC bombers could penetrate Soviet air defenses with high confidence. Extensive deployment of more capable Soviet SAMs, beginning in 1958, made this task more difficult, forcing SAC to shift its efforts to low-altitude penetration. Nonetheless, U.S. expectations of bomber penetration remained at roughly 75 percent.[5]

The long-range ballistic missile promised to make SAGE (and, for that matter, PVO) highly vulnerable to attack. SAGE's direction centers relied for their survival on their air defense forces, and those forces were helpless against ICBMs. In response, hardened "super-combat centers" buried deep underground were considered and rejected: Even if such centers survived attack, the reasoning went, the system's infrastructure—its radars, telephone lines, SAM sites, and air bases—would remain vulnerable to Soviet missile strikes.[6] SAGE, for all its remarkable technical achievements, had been overtrumped by Soviet offensive technologies. The Soviet ICBM not only negated U.S. air defenses, but the task of developing a missile intercept capability, begun by the United States in the mid-1950s, proved to be technologically daunting when it came to complex engagements.

Even before the advent of Soviet ICBMs, however, there was official concern about the vulnerability of U.S. forces to Soviet surprise attack. Analysts at the RAND Corporation recognized early on that a highly vulnerable U.S. force was a poor deterrent—deterrence lay not in the gross size of the U.S. weapons inventory but in how much weaponry could be delivered on target under the most adverse circumstances. Although SAC possessed a large nuclear air force, it was concentrated on only a few dozen bases. A random readiness check at one SAC base in September 1956 found that no SAC bomber could be armed and airborne inside the six-hour maximum warning time expected in a surprise air attack (weapons were stored apart from the aircraft, in shelters that were not blast-hardened). One year later, SAC was found to be capable of "flushing" not more than 8 B-52s, 9 B-36s, and 117 B-47s—or about 8 percent of the total force—given two hours warning of attack.[7]

Why was SAC so unprepared to launch on actual warning of Soviet attack? First, SAC expected to receive lengthy, strategic warning of Soviet attack—for example, an international crisis, or indications of Soviet mobilization from U.S. intelligence—and having more than a few hours to respond rendered the vulnerability issue moot. Second, the Eisenhower administration's doctrine of massive retaliation contemplated a preemptive strike. President Eisenhower expected to launch SAC forces against the Soviet Union "as soon as he found out that Russian troops were on the move."[8] Soviet forces would be expected to suffer significant attrition from SAC bombardment. However, under such a scenario, Soviet bombers might have time to be armed, dispersed to secondary airfields, and launched on warning of U.S. attack.[9]

Eisenhower was not sanguine about the prospects for constraining retaliatory damage to U.S. society if the United States *did* preempt. The prospect of even a fraction of the Soviet nuclear arsenal landing on American soil gave U.S. leaders sufficient pause to render the likelihood of U.S. preemption low.[10] It is not the case, then, that in the 1950s vastly superior U.S. striking power was largely subject to self-restraint. It was dogged at every step by Soviet power, real or imagined. But by the end of the decade, SAC had developed a system of alerts sufficient for both retaliation and preemption, with a portion of the bomber force fueled, armed, and ready to fly at all times. The air defense system then reaching operational status also offered SAC better protection against the modest Soviet bomber threat, and improved U.S. intelligence showed that as late as 1961 the Soviets had no more than four ICBMs operational.[11]

As Soviet strategic forces grew, the credibility of U.S. "extended" deterrent threats on behalf of NATO, particularly threats advocating a nuclear response to conventional attacks, also came increasingly into question. When the Kennedy administration came into office in 1961, it began to urge a shift in NATO doctrine from massive retaliation to "flexible response"—meeting conventional attack with conventional defense, using nuclear weapons only *in extremis* or in response to Soviet nuclear use.[12]

The Kennedy administration also changed U.S. nuclear declaratory policy to second-strike rideout (although U.S. bombers on alert would continue to be launched on warning under positive control), accompanied by explicit counterforce retaliatory targeting and city avoidance (the former not new to operational planners and the latter not necessarily feasible, given the proximity of many military targets to civilian facilities).[13] In 1964, declaratory policy was further refined to "assured destruction/damage limitation." Under this policy, U.S. strategic forces were to be capable of "assuring" destruction of 20 to 25 percent of the Soviet population and 50 percent of its industry, even after absorbing a Soviet first strike. Four hundred equivalent megatons (EMT), delivered and detonated, was one force requirement, an amount calculated to be on the "flat of the damage curve" (doubling the megatonnage would raise damage levels by only a few percent at best). Assured destruction was the peacetime deterrent threat. Damage limitation was the corollary wartime requirement that U.S. forces striking the Soviet Union in retaliation reduce as far as possible the Soviet Union's remaining capacity for war.[14]

In 1955, the Soviet Union initiated research on ballistic missile defense (BMD), about the same time as the United States did; in 1961, it began to deploy a rudimentary ABM system around Leningrad. However, in 1962 construction was halted, at about the same time that ground was broken on a different system, also pegged as an ABM, around Moscow. In 1963, the never-completed Leningrad system was dismantled, and construction started on a third system, initially thought to be an ABM as well, which ran northeastward in a line from the Estonian city of Tallinn. The exact nature of the "Tallinn System," with its very long-range SA-5 interceptors, remained a point of controversy within the U.S. defense and intelligence communities for several years. Ultimately it was classified as an air defense system, but the United States continued to monitor it closely for evidence of "upgrade" to ABM status, especially after signature of the ABM Treaty.[15]

U.S. development of its BMD carried over from the Eisenhower administration. President Kennedy inherited a program, Nike Zeus, that his predecessor's Science Advisory Committee had found to be too slow, too vulnerable to attack, and too susceptible to saturation (by warheads or decoys, which it could not differentiate between).[16] By 1965, the U.S. Army had developed phased-array radars (scanned electronically and built into blockhouses ten times "harder" than the old Zeus radars), and new, faster Sprint interceptors in a system dubbed Nike-X. In 1966, the Joint Chiefs of Staff unanimously recommended deploying this system to defend U.S. territory against Soviet missile attack (that is, as an "area defense").

Defense Secretary Robert McNamara waged an internal battle against Nike-X deployment. Freedman assesses his reasoning, and that of his opponents, as follows:

> To McNamara ABM procurement appeared as a snare and a delusion; a truly effective ABM did not seem feasible, while the attempt to get one would stimulate an arms race. . . . It might, he thought, be possible to stabilize the [action-reaction phenomenon] by withholding a stimulus in the hope that a response would not then be forthcoming. . . . In the 1960s this specifically meant eschewing the deployment of ABMs. To the military this represented extremely dangerous reasoning, placing far too much trust in Soviet restraint. They believed that if there were strategic opportunities to be exploited then it was incumbent upon the U.S. to exploit them before the Soviet Union. . . . The ABM therefore became a deeply symbolic issue, bringing to a head fundamentally opposite views concerning the most appropriate form of arms race behavior for the U.S.[17]

In other words, ABM was (and is) not only a technology issue, but also one that affects basic U.S. national strategy and is affected by beliefs about the nature of the Soviet Union and about the desirability and utility of U.S.-Soviet cooperation.

In late 1966, McNamara used technological as well as political arguments to fend off Nike-X deployment. He consulted the system's prime contractors, who told him that Nike-X was not yet ready for deployment. He elicited the Joint Chiefs' assent in postponing a deployment decision until the Soviet Union could be sounded out on the issue of limiting ABMs. In January 1967, he arranged to have President Johnson meet with all past and present presidential science advisors and directors of defense research and engineering, who informed Johnson that the Nike-X system was incapable of defending against a concerted Soviet missile attack.[18]

McNamara nonetheless remained under pressure to respond to the Moscow ABM deployment (construction of which had gained momentum in 1965 after languishing for several years) and to the rapid growth of Soviet strategic missile forces as the third generation of Soviet ICBMs came on-line. In 1966, over 300 silos for these missiles (the SS-9 and SS-11) became operational, and the pace continued into 1967. U.S. missile deployments, on land and at sea, levelled off in 1967, with no future increases planned. Johnson, who in the 1950s had exploited the "missile gap" issue for political purposes, did not want to contend with a "gap" of his own in the 1968 election.[19]

McNamara supported a solution on the offensive side, namely, the development of multiple, independently targetable reentry vehicles (MIRVs). Without changing the number of operational missiles, a MIRV system could multiply U.S. ability to hit retaliatory targets in the Soviet Union severalfold. The additional warheads also could be used to overload Soviet missile defenses. But to avoid having to respond with ABM deployment as well, McNamara needed some sign of Soviet willingness to restrain ABMs. When, at the impromptu Glassboro Summit of June 1967, Soviet Premier Kosygin seemed to rule out that option, that potential check to U.S. ABM deployment was removed. (Ironically, in 1968 work on the Moscow ABM system halted anyway, most likely because it was judged to be technically deficient.)[20]

In September, bowing to domestic pressure, McNamara announced a decision to deploy area defenses based on Nike-X technology and dubbed "Sentinel," ostensibly intended to guard against any future Chinese missile threat. Sizing Sentinel to the Chinese threat was McNamara's way of giving in to ABM proponents while trying to minimize the impact he thought it would have on the U.S.-Soviet arms race. Critics of McNamara's decision were inclined to be more cynical, noting that Sentinel could be expanded readily into a much "thicker," Soviet-oriented system. The Joint Chiefs supported it largely because it was expandable; other U.S. officials saw in that potential the necessary leverage to get the Soviet Union to the arms control bargaining table.[21]

The Sentinel decision raised the curtain on the first great U.S. ABM debate. The Johnson administration had expected some protest by cities that stood to be less well protected by the Sentinel system; it did not expect protest from cities that stood to be best protected, such as Chicago, Boston, Detroit, and Seattle. But Sentinel would have placed relatively large numbers of nuclear weapons near those cities (the Spartan interceptor, the principal means of city defense, carried a five megaton warhead), and engendered a "not in my backyard" response familiar to anyone responsible for siting potentially risky public facilities (for example, prisons, nuclear power plants, toxic waste incinerators).[22]

In early 1969, the Nixon administration reevaluated the Sentinel system and in March reoriented it toward defense of ICBM silos, renaming it "Safeguard." Congress, in having to reauthorize deployment, reignited the debate over the wisdom of deploying ABM. Some critics contended that the Safeguard system, which used the same equipment as Sentinel but in somewhat different proportions, was still vulnerable to defense suppression attacks from Soviet multiple-warhead missiles. The detonation of Safeguard's own nuclear interceptors could blind its crucial guidance radars (the interceptors were command guided from the ground and had no independent means of homing in on their targets). If its own interceptors' detonations did not blind its radars, then Soviet precursor bursts surely would. Those deficiencies are now widely acknowledged to be valid, and they provide a major impetus for current U.S. BMD research to develop non-nuclear interceptors.[23]

The political storm over Sentinel and Safeguard, and the technical criticism, began to turn Congressional opinion against ABM. The souring situation in Vietnam also contributed to Congress's tendency to be critical. In August 1969, the Senate authorized deployment of four Safeguard sites near Minuteman bases on a very close vote. One year later, the Senate refused to authorize phase two of the system, which involved deployment of four area defense sites, as well as construction of additional sites for Minuteman defense. By this time, the United States and the Soviet Union were actively engaged in negotiations to limit strategic nuclear arms. Those negotiations ultimately would provide the main rationale for continuing Congressional support of Safeguard, as a bargaining chip.[24]

The United States sought negotiations after it had completed its own strategic missile buildup; the Soviet Union's corresponding buildup was incomplete but moving ahead at full steam. The Soviet Union expressed interest but delayed serious negotiation until its own forces (both deployed and still in the pipeline) had roughly approached parity with those of the United States. Rough parity in land-based missiles was achieved in 1970. By 1972, continuing Soviet ICBM deployments exceeded those of the United States by 50 percent, although Soviet sea-based missile deployments lagged behind, and its bomber force remained at 1959 levels. When the SALT I agreements were signed in May 1972, the two countries had equivalent numbers of strategic nuclear delivery vehicles.[25]

Strategic parity was a prerequisite for serious arms control negotiations—neither side wished to negotiate from a position of inferiority. But parity was at least as discomforting for the United States as its initial sense of vulnerability had been fifteen years earlier. As Soviet nuclear capabilities grew, the proportion of U.S.-Soviet crises involving U.S. nuclear threats declined. Growing confidence in the nuclear "correlation of forces" influenced the decision of the 24th Soviet Party Congress to endorse SALT in April 1971, and SALT in turn ratified Soviet status as a superpower. That confidence also furthered Soviet activism in the Third World in the 1970s, to the eventual detriment of SALT.[26]

In May 1972, the United States and the Soviet Union signed the ABM Treaty and the Interim Agreement on strategic offensive ballistic missiles. The ABM Treaty limited both parties to two ABM sites apiece, with no more than 100 ABM launchers and interceptors per site, one of which could protect an ICBM silo deployment area and the second of which could protect the national capital. After the U.S. Congress refused to authorize funds for building a site outside Washington, D.C., the number of permissible sites was reduced (in a 1974 protocol to the Treaty) to one apiece, either at the ICBM silo deployment area or at the national capital. In early 1975, the United States deployed its system (the last remnant of Safeguard) near the Minuteman fields at Grand Forks Air Force Base (AFB), North Dakota. However, the system was deactivated by Congress within a year, on the grounds of low cost-effectiveness. Since a mere 100 U.S. ABM interceptors could do little to thwart several thousand Soviet warheads, there was no reason save troop training to keep Safeguard operational.[27]

The Soviet Union maintained its permitted ABM deployment at Moscow, but did not at first add to the sixty-four interceptors then in place. Indeed, for twelve years the system remained outwardly unchanged. In 1980, thirty-two interceptors were removed preparatory to a major system upgrade, which was within bounds of the ABM Treaty. Currently nearing operation are a new large phased-array radar (LPAR) and silos for short- and long-range ABM interceptors, a configuration much like that of the U.S. Safeguard system abandoned a decade ago. A chain of newly constructed early-warning LPARs now nearing completion will upgrade current capabilities and close existing gaps in Soviet missile early-warning coverage.[28]

The 1972 SALT I Interim Agreement limited the number of strategic missile launchers on both sides but did not limit the number of warheads per missile. Nor did it address the issue of missile accuracy, which in the late 1960s had been seen as an eventual threat to hardened military targets. The "window of vulnerability" argument soon began to take shape, spurred by the deployment of fourth-generation MIRVed Soviet ICBMs and by actions of the Committee on the Present Danger (CPD), which warned that a "limited" Soviet counterforce strike could destroy U.S. land-based missiles and leave a U.S. president to choose between "suicide or surrender"—an exchange of destroyed cities or no retaliation at all.[29] Prospective U.S. ICBM silo vulnerability to Soviet ICBMs also increased interest in developing a "hardsite" or "point" defense.

Hardsite defense had been the focus of U.S. BMD research since the ABM Treaty was signed, and in 1974 the U.S. Congress directed that that effort focus on basic technology development rather than on prototype system engineering. In 1976, Congress voted against allocating a modest amount of funds for research on area defense, and established a funding level for BMD (roughly $200 million in fiscal 1977 dollars) that remained relatively constant until fiscal 1982, when funding began to rise (by 48 percent or $130 million, which is not much by later SDI standards but was the first significant growth in over a decade).[30] U.S. BMD research remained focused on hardsite defense, however, with the Army struggling to keep pace with the latest proposed modes for basing the large, new MX ICBM.

The Army designed a conceptual Low Altitude Defense System (LOADS), to defend the MX in multiple protective shelters (MPS or the "racetrack") only to see it cancelled in favor of Closely Spaced Basing (CSB or "dense pack"). The Army then devised a conceptual defense for CSB, only to see this system rejected by Congress in late 1982.

POLICY AND RHETORIC, 1983–1987

Early Reagan administration rhetoric about the need to prevail in a nuclear conflict, and neo-Cold War rhetoric against the Soviet Union, elicited powerful public reactions in both the United States and Western Europe, giving rise to the Nuclear Freeze movement in the United States and to growing protests against deployment of inter-mediate-range nuclear missiles in Western Europe. Although initially committed to refrain from new arms talks until U.S. strength had been "rebuilt," due to these pressures the Reagan administration found itself back at the bargaining table, first on intermediate-range nuclear forces (INF) in November 1981, then for strategic arms-reduction talks (START) in mid-1982, even though its nuclear build-up had yet to pass the stage of moral rearmament. Few foresaw that President Reagan would seek to end-run his nuclear critics, not only the advocates of a Nuclear Freeze but also the U.S. Conference of Catholic Bishops, with an appeal to turn the existing structure of deterrence and arms control inside out.

Reagan had pondered the question for some time: It seemed to him the height of folly to leave the United States vulnerable to Soviet attack and not pursue development of U.S. defenses with the greatest urgency. After the 1980 election, he raised the issue of what would become SDI with then-Senator Harrison Schmitt. He was briefed in early 1982 by retired U.S. Army General Daniel Graham, former head of the Defense Intelligence Agency (DIA), on the pri-vate High Frontier group's proposal for a low-cost, low-tech missile defense using a combination of orbiting "missile trucks" and ground-based conventional interceptors. The following September, Edward Teller, codeveloper of the hydrogen bomb, briefed Reagan and White House aides on progress in developing the X-ray laser at Livermore National Laboratory.[31]

When Congress rejected dense pack in late 1982, Reagan then turned to a blue-ribbon panel, headed by former National Security

Advisor Brent Scowcroft, to evaluate the strategic nuclear modernization program and the role of MX within it.

In February 1983, two months before the Scowcroft Commission issued its pivotal report, Reagan met with the Joint Chiefs to discuss the question of active defense for the MX system. After a long, fruitless search for a survivable basing mode, the MX appeared to be headed for placement in ordinary Minuteman silos and concommitant extraordinary vulnerability. Reagan engaged the Joint Chiefs on strategic defense for half an hour, but not on the subject of hardsite defense, asking instead, in language that has since become familiar, whether it would not be better to save lives than to avenge them. Chief of Naval Operations Admiral James Watkins agreed that it would.[32]

All of this was prologue to a presidential speech on March 23, 1983, in which Reagan, without advance warning to the U.S. defense policy bureaucracy, NATO, or the Soviets, announced a major shift in U.S. policy. Reagan's "vision" was simple; its implementation was not: replace deterrence with defense; task science to undo what it had done forty years before.

First to be rendered impotent and obsolete by the President's speech was the report of a White House Science Council panel that earlier in 1983 concluded that third-generation nuclear weapons, including the X-ray laser, had little future in BMD and that also downplayed the ultimate military potential of "directed energy" technologies.[33]

Unaffected by Reagan's speech, the Scowcroft Commission recommended basing the MX in Minuteman silos, initiating the development of a small, single-warhead, mobile ICBM (the "Midgetman"), and devoting serious attention to arms control. It did not recommend defenses for MX, viewing BMD as a detriment to acceptance of MX by Congress. Rather, it deemphasized the importance of ICBM vulnerability in light of the survivability of other elements of U.S. strategic forces (manned bombers and submarines).[34] In so doing, it closed the window of vulnerability and highlighted that concept's fundamentally political origins: opened to promote the deployment of a survivably-based MX; closed when survivable basing, no longer feasible, made the concept a liability to deployment. In thus closing the window of vulnerability, the Scowcroft Commission removed the principal rationale for active defense of land-based missiles. However, since the administration's goal for BMD was about to be radically recast, this was not really a problem.

The March 23 speech transformed the terms of the Western debate over strategic nuclear weapons and policy. By seeming to seize the moral high ground, it also hastened the demise of the Nuclear Freeze movement. By legitimizing open discussion of all varieties of BMD for the first time in over a decade, it sent supporters of missile defense scrambling to define a technically supportable research program and a set of consistent program goals, both difficult tasks.

Two studies commissioned in the wake of the speech reached opposite conclusions. The Defense Systems Technology Study, headed by once-and-future NASA Director James Fletcher, pressed for research into exotic technologies that had long-term, high-value payoffs for population defense. The Future Security Strategy Study, headed by Pan-Heuristics Chairman Fred Hoffman, pressed for a BMD policy with a more immediate payoff, including defense of U.S. land-based missiles and defense in Western Europe against Soviet tactical missiles.[35] Although cautionary in part, both studies were far more positive on BMD than studies completed just one year earlier.

Nine months after Reagan's speech, the Strategic Defense Initiative was born. The debate over SDI is not just a debate about technology, nor is it only about appropriate U.S. arms race behavior. Moreover, since the first ABM debate ended, the framework for debate has changed. The defense-regulatory mechanism that McNamara sought has been in place since 1972. Whether one believes that the ABM Treaty and arms control in general have been good for the West depends on more than just the details of the particular bargains struck. It also depends, importantly, on whether one believes such bargains should be struck at all. The conservative wing of the U.S. political spectrum, for example, has never been happy with arms control, seeing it as a one-way accommodation that has held back development of U.S. weapons technology and "lulled" Western publics into thinking that the struggle between communism and capitalism can be transformed into a game with known and respected rules. While the West hews to the terms of various agreements, Moscow cheats; sometimes it is caught, sometimes not. Arms control, in short, is bad for the West. The ABM Treaty is the cornerstone of arms control. Therefore, the Treaty should go.[36]

Still, although the U.S. public is generally unaware that the country lacks an effective missile defense system, generally favors the development of a non-nuclear-based defense system, and generally does not trust the Soviet Union, it also tends not to want to dismantle the existing framework of arms control, a point on which NATO and the U.S. Congress are also quite sensitive. Congress has consistently admonished the Reagan administration to see that U.S. R&D programs, including SDI, are kept fully compliant with the Treaty. Congress's focus on hardsite defense can be viewed in that light.[37]

However, defense of strategic forces is not what the President or Defense Secretary Weinberger have in mind when they speak of SDI. Both stress that SDI's purpose is to protect people, not weapons, and both stress the morality of that position. As former presidential science advisor George Keyworth put it, "Protecting weapons . . . simply strengthens the doctrine of mutual assured destruction. Protecting people, on the other hand, holds out the promise of dramatic change."[38] Yet the moral superiority of the pro-SDI position is asserted both by those who would enhance deterrence as well as those who would supplement it.

Weinberger has argued both positions, viewing them as two different phases in the postulated transition to a defense-dominant world. But he tends to focus on the end result. SDI is "one of history's best chances to end the shadow and the fear of nuclear weapons"; a chance to be free of the "dogma" of "mutual assured vulnerability," which has kept U.S. strategic defense efforts shackled while Soviet programs surge ahead. Weinberger is intolerant of SDI's critics, recently characterizing them as lacking both the moral fiber to defend democracy and the moral vision to distinguish it from tyranny.[39] His views on arms control and the ABM Treaty were not shaped solely in response to Reagan's March 1983 speech, however; his dislike is apparent in a number of pre-1983 statements.[40]

Undersecretary Iklé has also been a vocal critic of modern strategic doctrine on behalf of SDI. Shortly after Reagan's speech, Iklé characterized U.S. nuclear policy in the 1970s as the product of a "nightmare" view that "mankind must remain locked into permanent hostile confrontation of missile forces poised for instant retaliation . . . ," a view that "implicitly accepts a world of nations frozen into an evil symmetry. . . ." He contrasts this view with a "vision" that "searches for ways to stop a nuclear attack, rather than relying exclusively on the threat of revenge," a vision that therefore "offers hope."[41] A 1985 article elaborates, describing the "consensual vulnerability theory" of nuclear deterrence as being the strategic equivalent of the "flat earth"—a rarefied theoretical construct divorced from reality, in this case from the reality of Soviet strategy and behavior. Its acceptance by the United States reflects "not a state of nature but a state of mind." In other words, society's vulnerability to nuclear weapons is not a condition of the nuclear age, but a conscious choice. It is not the destructive power of nuclear weapons that has made defense against them difficult but the unwillingness of past administrations to pursue a solution aggressively, in the at best naive hope that the Soviet Union "would not want to overtake us in nuclear offensive forces, much less seek a capability for destroying most of our deterrent strength." The ABM Treaty "became the apotheosis of this belief. . . ." However, the effort to lock in consensual vulnerability failed because "our arms control efforts . . . failed to halt the competition in counterforce capabilities." The net result is a "quivering balance" of nuclear capability. A "stable equilibrium of mutual restraint is psychologically incompatible" with such a "constant threat of reciprocal annihilation [and] unremitting determination to deny [the other side] escape from this grip of terror." Offense-based deterrence, in short, makes peace a fragile commodity and arms control a doomed effort. The Soviets, according to Iklé, have a choice: either to cooperate in a joint transition to the "new order" of defense dominance or attempt to block U.S. progress along the "harder road" of competitive transition.[42]

As a third example, the 1983 Hoffman Panel report stressed that emphasizing defenses could reduce U.S. "reliance on threats of massive destruction that are increasingly hollow and morally unacceptable." This statement implies both that the United States relies on countervalue threats ("massive destruction" of Soviet cities and industry) to implement deterrence, and that peace will be increasingly hard to secure because the United States will be seen as being unlikely to implement its ("hollow") deterrent threats for fear of Soviet reprisal.[43] Paul Nitze voiced similar criticisms in the mid-1970s, although the preferred solution then was not defense but greater counterforce capability.[44]

In 1985 Senate testimony, Hoffman charged not that current U.S. policy was immoral but, that if given the opportunity, arms controllers would make it so. Mutual assured destruction (MAD), according to Hoffman, is the "orthodoxy reflected in the SALT process" and the "strategic foundation" underlying the ABM Treaty, as well as a "largely unconscious dogma dominating the media discussions of nuclear strategy, SDI, and arms agreements." Advocates of a MAD national strategy consign nuclear weapons to deterrence of Soviet nuclear use "through the threat of massive and indiscriminate attacks on cities," the morality and prudence of which were being questioned by "a broad and increasing segment of the public." For Hoffman, the moral issue lies not in offense-based deterrence per se but in this postulated "indiscriminate" targeting of cities. Thus, "[t]he relevant question for the foreseeable future is not whether defenses should replace offensive weapons but whether . . . a combination of militarily effective and discriminating offense and defenses will better meet our strategic requirements for deterrence and limiting damage."[45] Hoffman thus may be seen to represent the deterrence enhancing wing of the pro-SDI community.

The preceding reflects a sort of revisionist history of the past fifteen years that slides by most of the doctrinal debates of a period that has seen the convergence of declaratory (that is, public) policy and operational doctrine toward establishment of a common counterforce, warfighting orientation. Defense secretaries from James Schlesinger (1973) forward have sought to reduce the size of a minimum U.S. nuclear response to a level that presidents could at least conceive of as limited, reasoning that if war plans contained only "splendid," all-out responses, attempts to contain the scope of a nuclear war would be impossible. Whether a nuclear war could be limited is of course unknown, but the probability of limited nuclear war is low for a host of reasons, not the least of which is Soviet nuclear strategy. Nonetheless, since 1974 U.S. doctrine has evolved from the "limited nuclear options" of NSDM 242, to the "countervailing strategy" of PD 59, to the Reagan administration's call in NSDD 13 to "prevail," if need be in a "protracted" conflict.[46]

These are not "assured destruction" strategies; U.S. strategy has, if anything, been moving away for over twenty years from the kinds of city targeting criticized by Hoffman and others, leaving MAD as a force planning benchmark and an underlying condition of the nuclear age. The limited options and war-fighting strategies developed in part to compensate for a perceived decrease in the credibility of U.S. deterrent threats in the face of Soviet assured destruction capabilities and in part to avoid activation of those capabilities in the event of war. Defenses designed to enhance deterrence are one logical next step in this progression, and defenses to replace deterrence are the (very precipitous) step after that. However, technology is the servant of offense and defense alike, offers no sure exit from the nuclear dilemma, and promises to make a concerted effort to find one both protracted and frustrating.

SDI proponents' claims to moral superiority are also hard to credit. Not until a defense reached 75 percent effectiveness against a Soviet second strike would it even begin to undermine Soviet capacity to destroy the United States as a functioning society, given current numbers of deployed warheads; not until effectiveness against a Soviet *first strike* exceeded 99 percent would the job approach completion.[47] Until it is efficient enough to absorb all but a handful of incoming warheads, strategic defense would work best to break up closely timed counterforce attacks. City defense would remain the hardest defense task, and city attack the easiest offense task. Short of near-perfect performance levels, strategic defenses not only would fail to overcome the *condition* of assured destruction, they would make it the default nuclear strategy. In short, it is "reckless," as former Defense Secretary Schlesinger writes, to base justifications for strategic defense "on assertions regarding the 'immorality' of deterrence," because "for the balance of our days, the security of the Western world will continue to rest on deterrence."[48]

Thomas Schelling offers a final corrective. "The correct name...," he writes, "is not 'assured mutual destruction,' but 'assured *capability* for mutual destruction,' the difference being that the capability does not have to be ineluctably exercised at the outbreak of even an intercontinental nuclear war." Schelling argues that a "retaliatory targeting" strategy backed by such a capability could have built-in options to withhold certain weapons or avoid certain targets, as well as limited options, the same as a counterforce strategy. It could be less escalation-prone than counterforce and it would present fewer incentives to preempt (an oil refinery is not a critically time-urgent target).[49] Offense-based deterrence thus could conceivably thread a path between war-fighting doctrines and a slaughter of the innocents. But this strategy, like any strategy to constrain nuclear war, would depend for its success on the Soviet response, and Soviet cooperation in limiting U.S. damage is not what Soviet doctrine heretofore has been about.

SOVIET NUCLEAR STRATEGY
AND ARMS CONTROL

Soviet strategy is designed to deter nuclear war by denying victory to the enemy and by limiting damage to the Soviet homeland to the greatest extent possible. It is a strictly military strategy—Soviet strategists wear the uniform of the General Staff; none are civilians. On the other hand, the doctrinal parameters guiding the development and operation of Soviet nuclear forces are set by the civilian political leadership. Those leaders take pains to exercise close control over nuclear forces; and they fully appreciate that nuclear war would be an unmitigated disaster for the Soviet Union. That reality has been recognized since at least the time of Khrushchev; it was explicitly incorporated into military doctrine following Leonid Brezhnev's 1977 speech at Tula.[50]

Michael MccGwire summarizes this juxtaposition of preparedness and fear as follows: " . . . [The Soviets] place a very high priority on avoiding [a world] war . . . , but should it prove inescapable, their objective is to prevail, or at least not to lose. It is this latter very demanding requirement, and not deterrence doctrine, that determines Soviet force posture and operational requirements."[51] That is, Soviet leaders acknowledge the reality of "assured destruction," but they do not like it and instruct the military to plan to mitigate it should war come.[52] According to Stephen Meyer, those plans advocate no single response but one that varies according to circumstances. These circumstances are doctrinally defined as: (1) a surprise attack; (2) a war arising from a crisis in which the attack is anticipated; (3) escalation from a conventional war; and, as lesser contingencies: (4) an accidental war; and (5) escalation from a local war (for example, in the Middle East). There are three "operational responses" to these contingencies: preemption (launch on strategic warning—indications of imminent enemy attack); launch under attack (that is, on tactical warning), and second-strike ride-out (for "assured retaliation," the default option, least desirable but most robust and thus the most likely to be available, given what can go wrong in wartime). To these Meyer adds, "the never-mentioned [in Soviet writings] Soviet disarming first strike"—the bolt from the blue that historically has been a major U.S. concern. It is an option that "the Soviet military has been unable to present . . . as an attainable goal—in part because of technical uncertainty but largely because the Americans won't hold still . . .''; that is, the United States might launch not only bombers but missiles on tactical attack warning.[53]

Soviet planning for nuclear war does not parallel the West's propensity to view the issue in terms of bargaining and game theory. Soviet planners are more keenly interested in decisively crushing the enemy's ability to fight; there is no indication of Soviet interest in intrawar bargaining, limited aims, or early war termination, particularly in the context of an intercontinental nuclear war.[54] As Meyer puts it:

> Once one moves to strikes on the homeland, Soviet abilities to discern the scale and purpose of the attack decrease rapidly. Thus the Soviets' strategic culture and history are likely to be the primary determinants of their immediate reactions. Everything in the Soviet experience points to the assumption that the attack has a single strategic purpose: to destroy the Soviet state. . . . At that point, the Soviets risk far less by assuming that an apparently limited nuclear strike is the first wave of a massive strike than by waiting to see what actually happens.[55]

Fritz Ermarth observes that "both institutionally and operationally, Soviet intercontinental strike forces are an outgrowth and extension of forces initially developed to cover peripheral targets." From a Soviet military perspective, the U.S. homeland is the strategic rear area of the Soviet Union's most important peripheral theater, Europe, so that contingencies and concepts for theater and strategic warfare, which are treated distinctly in U.S. discourse, tend to be melded in Soviet thinking and planning.[56]

Since about the mid-1960s Soviet doctrine regarding theater war has been evolving away from early and decisive use of nuclear weapons, leaving open the possibility of early nuclear use but stressing the *deterrent* value of theater nuclear forces and the use of conventional forces against NATO nuclear potential at the earliest possible point in a war. Modernization of Soviet conventional forces earmarked for the European theater over the past fifteen years is consistent with such a strategy. Theater nuclear forces themselves have a preemptive role to play, a role that is linked to the political leadership's assessments of enemy movement toward nuclear use. Once weapons are released, "Soviet military strategy calls for preemptive nuclear strikes against targets distributed across the entire depth of the theater.[57]

Strategic defense has a role to play in Soviet strategy as a damage-limiting element, taking some of the damage-limiting burden off the offense either by absorbing U.S. retaliation or by buying time for Soviet forces to switch to the offensive in the event of U.S. attack. But offense and defense are "inexorably linked" in Soviet military thinking; that rules out adherence to a purely defensive operational strategy.[58]

If Soviet military planning "adheres to no known yardsticks of strategic adequacy" comparable to the U.S. assured destruction concept, as Benjamin Lambeth and other analysts contend, then the growth of Soviet arms is constrained "only by domestic economic and technological resources, U.S. forebearance, and the formal protocols of negotiated arms limitation agreements."[59] U.S. opportunities to influence the size and shape of the Soviet military (short of war) are to be found in the second and third factors listed, namely, tacit and formal arms control.

Arms control may appeal to Soviet leaders on several grounds: political, operational/strategic, technical, and economic. As noted earlier, SALT ratified Soviet status as a superpower, thus granting it "political parity" with the United States.

In operational/strategic terms, SALT I and particularly the ABM Treaty have provided a stable environment for military planning and supported Soviet counterforce targeting options, thereby contributing to Soviet concepts of damage limitation. Stephen Meyer and Peter Almquist conclude that in the 1970s and 1980s, the Soviet Union had maximized its preemptive counterforce potential (within SALT constraints) *at the expense of* simultaneously maximizing it for the United States. That is, the options exercised had the effect of boosting U.S. counterforce potential even as they increased that of Soviet forces. Given Soviet emphasis on preemption and launch under attack, that tradeoff was deemed acceptable.[60] The ABM Treaty is consistent with such a strategy.

With respect to technology, analysts of Soviet military policymaking suggest that Moscow became keen to limit ABMs once it had decided that its missile defense technology would lag behind that of the United States, as would its relative ability to penetrate U.S. missile defenses; unconstrained competition would find the Soviet Union worse off both going and coming. The Treaty also helped to check "the United States' capacity for dictating the evolution of the arms race," a continuing motivation both for Soviet support of the Treaty and for its opposition to SDI.[61]

Soviet economic interest in arms control stems from the escalating cost of sophisticated arms and from the problems plaguing the Soviet economy. The prospective cost of countering or matching SDI and other U.S. military technology efforts may be very high—not high enough to induce Soviet systemic collapse, but perhaps high enough to stymie the long-term economic reforms that appear to be key to the current Gorbachev program.[62]

In short, Soviet leaders and to some extent the Soviet military have clear interests in arms control. These differ from those of the West, which has an interest in limiting Soviet military power, building stable relations, and encouraging the reallocation of Soviet resources to non-military ends. To reach agreement, both sides make tradeoffs, cut the best deal they can, and closely monitor their "partner's" subsequent compliance. Compliance has been a sore point in recent years, because of Soviet practices and because of the ascendancy of hardline conservatives in the U.S. defense/arms control bureaucracy, which demonstrate again that arms control is a matter not only of military technology but of political will.

The 1972 ABM Treaty has proven to be as durable as any arms control agreement in modern history, but time, technology, and political shifts have taken their toll on it. Chapter 2 examines the costs and benefits of the Treaty from the U.S. perspective and reviews some problems that would need to be addressed if the Treaty is to remain an effective instrument of arms control.

NOTES

1. "No supportable estimate" could be made of the air threat to Europe based on then-available intelligence data. Since the Soviet TU-4 Bull was the aircraft thought to pose a threat to parts of the continental United States as well, no supportable threat estimate could be made for U.S. territory, either. U.S., Joint Intelligence Committee, *Soviet Atomic Capabilities Against Central Europe,* Joint Intelligence Committee Series, Rept. 558/109, 17 Apr. 1952. Declassified 1976. Published by the *Declassified Document Reference Service* (Washington, D.C.: Carrollton Press, Inc., 1976), Document 249B (hereafter cited as *DDRS* [year] [document no.]).

2. *New York Times* (hereafter *NYT*), 22 Jan. 1950, p. 17; *NYT* 14 Apr. 1950, pp. 1–2. However, civil defense was always assigned relatively low priority in continental defense policy, and by 1951 public apathy had set in: The House of Representatives that year cut the Truman administration's civil defense budget request from $535 million to $65 million with-

out public protest. *NYT*, 2 Sept. 1951, p. 8E. Since then civil defense has enjoyed brief intervals of public and government attention, but never consistent, strong public support.

3. U.S. National Security Council, *Continental Defense,* Rept. 5408 (Washington, D.C.: National Security Council, 11 Feb. 1954), pp. 7, 10. Declassified 24 June 1975. *DDRS*, 1975, 82A.

4. *NYT*, 4 Apr. 1954, p. 7. CONAD combined the twenty-five wings (3,000 aircraft) of the Air Force's Air Defense Command with the Army's surface-to-air missile batteries. CONAD became the North American Air Defense Command (NORAD), a joint U.S.-Canadian command, in June 1955. NORAD became a U.S. unified command three years later. *NYT*, 3 May 1959, p. 1. John F. Jacobs, "SAGE Overview," *Annals of the History of Computing* (Oct. 1983): 328–30. SAGE was originally planned to have 9 combat operations centers and 40 direction centers, but prohibitive costs reduced its final deployed size. For initial plans, see a briefing for the Senate Armed Services Committee, January 1955 (briefer's affiliation not indicated). Declassified 27 Nov. 1978. *DDRS*, 1981, 56A. Air defense deployment numbers are from U.S., National Security Council, *Continental Defense*, Report prepared by John Rubel (Washington, D.C.: National Security Council, 15 Sept. 1960). Declassified 14 Feb. 1984. *DDRS*, 1984, 2001. See also *Statement of Defense Secretary Robert McNamara Before the House Subcommittee on Defense Appropriations on the . . . FY 1964 Defense Budget*, 6 Feb. 1963, p. 51. Declassified 25 June 1975. *DDRS*, 1975, 150C.

5. David Holloway, *The Soviet Union and the Arms Race* (New Haven: Yale University Press, 1983), pp. 15–23; Robert P. Berman, *Soviet Air Power in Transition* (Washington, D.C.: Brookings Institution, 1978), pp. 15, 17, 19, 25–26; Robert P. Berman and John C. Baker, *Soviet Strategic Forces: Requirements and Responses* (Washington, D.C.: Brookings Institution, 1982), pp. 41–50; S.W.B. Menaul, "Air Defence of the Homeland," in Ray Bonds, ed., *The Soviet War Machine* (New York: Chartwell Books, 1976), p. 52; Harriet Scott and William Scott, *The Armed Forces of the USSR*, 2d ed. (Boulder, Colo.: Westview Press, 1981), p. 147. For estimated penetration rates, see Vann Howard Van Diepen, "Strategic Force Survivability and the Soviet Union" (M.A. thesis, Department of Political Science, Massachusetts Institute of Technology, 1983), pp. 210–11.

6. U.S., National Security Council, *Continental Defense,* note 4.

7. See Robert Cutler, "SAC Concentration in the U.S. and Reaction Time," Memorandum for the President, 25 Oct. 1957. Declassified 6 Apr. 1979. *DDRS*, 1984, 713. See also Andrew Goodpaster, Memorandum of Conversation, White House Briefing by Robert C. Sprague, 7 Nov. 1957. Declassified 8 Feb. 1979. *DDRS*, 1979, 331A. See also Fred Kaplan, *Wizards of Armageddon* (New York: Simon & Schuster, 1983), pp. 132–33.

8. David A. Rosenberg, "The Origins of Overkill: Nuclear Weapons and American Strategy, 1945-1960," in Steven E. Miller, ed., *Strategy and Nuclear Deterrence* (Princeton, N.J.: Princeton University Press, 1984), pp. 143-45, 152. Meyer notes that a U.S. surprise first strike was "the central doctrinal planning contingency for the Soviet military during the 1950s and through the mid-1960s." Nonetheless, neither Soviet bombers nor missiles were kept on alert during this period. Their warheads were stored in depots at some remove from forces and guarded by KGB troops. Stephen M. Meyer, "Soviet Perspectives on the Paths to Nuclear War," in Graham Allison, Albert Carnesale, and Joseph Nye, eds., *Hawks, Doves, and Owls. An Agenda for Avoiding Nuclear War* (New York: Norton, 1985), pp. 173, 175.

9. Richard Betts, "A Nuclear Golden Age? The Balance Before Parity," *International Security* (Winter 1986-87): 14.

10. Ibid., pp. 24-25, 28. Contrast with the assessment of Fred Hoffman, "The SDI in U.S. Nuclear Strategy," *International Security* (Summer 1985): 21. "The Soviets know that they can live under conditions of U.S. superiority without any serious fear of U.S. aggression because they have done so in the past. . . ."

11. *NYT*, 11 Nov. 1957, p. 12; *NYT*, 19 Nov. 1957, p. 1; *NYT*, 18 Dec. 1957, p. 1. Desmond Ball, *Politics and Force Levels: The Strategic Missile Program of the Kennedy Administration* (Los Angeles: University of California Press, 1980), p. 46; Daniel Ellsberg, "On Shaping the End of the Thermonuclear Age," *Transition* 4, no. 3 (November 1979): 2.

12. John Lewis Gaddis, *Strategies of Containment* (Oxford: Oxford University Press, 1982), p. 198-236.

13. Lawrence Freedman, *The Evolution of Nuclear Strategy* (New York: St. Martin's, 1983), pp. 234-44. Under a doctrine of "city avoidance," urban areas would not be deliberately targeted with nuclear weapons. A "second strike rideout" policy would not launch retaliatory missiles until an enemy attack had run its course; contrast with "launch on warning," that is, launch before incoming warheads detonate. Under "positive control launch," U.S. bombers fly to prearranged holding points from which they would not proceed to targets without receiving further orders.

14. Alain Enthoven and K. Wayne Smith, *How Much Is Enough?* (New York: Harper & Row, 1971), pp. 174-75, 207. The assured destruction threat was spelled out most succinctly by Thomas C. Schelling, *Arms and Influence* (New Haven: Yale University Press, 1966), pp. 22 ff. McNamara was heavily influenced to shift his policy away from counterforce by the damage limiting studies directed by Air Force General Glenn Kent. See Kaplan, *Wizards*, note 7, pp. 320 ff.

15. John Newhouse, *Cold Dawn: The Story of SALT* (New York: Holt, Rinehart & Winston, 1973), pp. 74-75; Lawrence Freedman, *U.S. Intelligence*

and the Soviet Strategic Threat, 2d ed. (Princeton, N.J.: Princeton University Press, 1986), pp. 88, 91.

16. U.S., Congress, House, Committee on Government Operations, *Organization and Management of Missile Programs*, 86th Cong., 1st sess., H. Rept. 1121, 2 Sept., 1959; U.S., White House, President's Science Advisory Committee, *Report of the Anti-ICBM Panel*, 21 May 1959. Declassified 4 Nov. 1980. *DDRS*, 1981, 181A.

17. Freedman, *U.S. Intelligence*, note 15, p. 83.

18. Enthoven and Smith, *How Much is Enough?* note 14, p. 187; Newhouse, *Cold Dawn*, note 15, pp. 85–6, 89.

19. Ibid., p. 84. The U.S. missile deployment rate peaked in 1964, when the United States put over 400 Minuteman ICBMs in the ground and launched 12 strategic ballistic missile submarines carrying 192 Polaris missiles. By 1967, Minuteman and Polaris launchers had stabilized at 1,000 and 656, respectively. As the United States finished its strategic missile buildup, the Soviet Union was just getting started. In 1963–64, the Soviet Union had roughly 200 heavy bombers; 32 submarines (two-thirds diesel-powered) with 3 short-range missiles apiece, which had to be launched with the boat surfaced; 130 second-generation SS-7 ICBMs in relatively soft horizontal shelters or "coffins;" and initial deployment of SS-7s in hardened silo launchers. Between 1964 and early 1966, the Soviet Union deployed only 50 additional SS-7 and SS-8 ICBMs, or 25 per year. However, between 1966 and 1969 the Soviet Union deployed an average of 250 ICBMs per year, and between 1969 and 1972 deployed over 100 modern sea-launched ballistic missiles per year aboard nuclear submarines roughly equivalent to U.S. Polaris boats. See Van Diepen, "Survivability," note 5, pp. 188, 192; U.S. Air Force, Strategic Air Command, Office of the Historian, *Development of SAC, 1946–1976* (Washington, D.C.: U.S. Government Printing Office, 1977); and International Institute for Strategic Studies, (thereafter IISS), *The Military Balance, 1974–75.* (London: IISS, 1974), p. 75.

20. Enthoven and Smith, *How Much is Enough?* note 14, pp. 185–94; Newhouse, *Cold Dawn*, note 15, pp. 84–95. On the MIRV program, see Ted Greenwood, *Making the MIRV* (Cambridge, Mass.: Ballinger Publishing Company, 1975), esp. pp. 49–50, 165.

21. Enthoven and Smith, *How Much Is Enough?* note 14, p. 193.

22. *Facts on File, 1969* (New York: Facts on File, Inc.), pp. 20, 215, 255, 418, 865; and Benson D. Adams, *Ballistic Missile Defense* (New York: American Elsevier, 1971), pp. 185–91.

23. See Steven Weinberg, "What Does Safeguard Safeguard," and Hans Bethe, "Countermeasures to ABM Systems," in Abraham Chayes and Jerome Wiesner, eds., *ABM: An Evaluation of the Decision to Deploy an Antiballistic Missile System* (New York: Signet, 1969), pp. 84ff, 130ff.

24. The Safeguard phase two prohibition is contained in U.S. Public Law 91-441, section 402, and cited in U.S., Congress, Senate, Committee on

Appropriations, *Fiscal 1971 Appropriations for the Department of Defense,* 91st Cong., 2d sess., S. Rept. 91-1392, 3 Dec. 1970, p. 24. The Senate Armed Services Committee justified its position (and overrode House opposition) on the grounds that the United States did not need anti-China area defenses. See U.S., Congress, Senate, Armed Services Committee, *Authorizing Appropriations for Fiscal Year 1971 for Military Procurement, R&D, and for Construction of . . . Safeguard. . . .* 91st Cong., 2d sess., S. Rept. 91-1016, 14 July 1970, p. 19.

25. Holloway, *The Soviet Union,* note 5, p. 59.

26. On nuclear diplomacy, see Barry M. Blechman and Stephen Kaplan, *Force Without War* (Washington, D.C.: Brookings Institution 1978), pp. 47-49, 127-29, 547-53. They emphasize that the *outcomes* of international crises in which U.S. forces were involved did not become any less favorable for the West as Soviet nuclear capabilities developed. On the other hand, the relative *frequency* with which strategic forces were involved in East-West crises declined steeply over time: from 1 in 5 crises, 1946-1948; to 1 in 6, 1949-1955; 1 in 15, 1956-1965; and 1 in 25, 1966-1975. Such a decline suggests at least increased caution in the use of strategic forces as a tool of crisis management, and indicates that the shift in the "correlation of forces" had some of the inhibiting effect on U.S. behavior that Soviet leaders and commentators ascribed to it. See Holloway, *The Soviet Union,* note 5, pp. 48 ff, 88-92; and Raymond Garthoff, *Detente and Confrontation* (Washington, D.C.: Brookings Institution, 1985), pp. 42-46, 53 ff.

27. On Safeguard termination, see U.S., Congress, Senate, Armed Services Committee, *Hearings on Military Procurement, Fiscal 1977.* Testimony of General Robert C. Marshall, 94th Cong., 2d sess., 30 Mar. 1976, pp. 6684-85, 6717; for Congressional reasoning on termination, see U.S., Congress, House, Appropriations Committee, *DOD Appropriations Bill, 1976,* 94th Cong., 1st sess. H. Rept. 94-517, 25 Sept. 1975, pp. 270-71.

28. IISS, *The Military Balance, 1985-86* (London: IISS, 1985), pp. 16, 164; William Durch and Peter Almquist, "The East-West Military Balance," in Barry Blechman and Edward Luttwak, eds., *International Security Yearbook, 1984-85* (Boulder, Colo.: Westview Press, 1985), pp. 40-43; U.S. Department of Defense, *Soviet Military Power, 1986* (hereafter *SMP*) (Washington, D.C.: U.S. Government Printing Office, 1986), pp. 43, 54, 57; and Carl Jacobsen, "BMD: The Evolution of Soviet Concepts, Research and Deployment," in David R. Jones, ed., *Soviet Armed Forces Review Annual, 1977* (Gulf Breeze, Fla.: Academic International Press, 1977), pp. 171 ff.

29. In a 1980 pamphlet, the Committee on the Present Danger (CPD) warned that the United States faced "a world crisis caused by the Soviet Union's lunge for dominance . . . a program of expansionism even more ambitious than that of Hitler. . . ." CPD, *The 1980 Crisis and What We Should Do*

About It (Washington, D.C.: CPD, January 22, 1980). In 1981, many prominent CPD members joined the Reagan administration. In October 1981, the administration unveiled its Strategic Modernization Program. In announcing the program, President Reagan stated that, "A window of vulnerability is opening, one that would jeopardize not just our hopes for serious productive arms negotiations, but our hopes for peace and freedom." ("Background Statement from White House on MX Missile and B-1 Bomber and "Transcript of Remarks by the President" in *NYT*, 3 Oct. 1981, p. 12.)

30.   U.S., Congress, Senate, Committee on Armed Services *DOD Authorization for Fiscal Year 1975*, Conference Report, 93d Cong., 2d sess., S. Rept. 93-1038, July 1974, p. 31; *idem., DOD Authorization for Fiscal Year 1977*, Conference Report, 94th Cong., 2d sess., S. Rept. 94-1004, June 1976, p. 33; and generally for fiscal years 1979-1984 statements on "Ballistic Missile Defense," U.S. Arms Control and Disarmament Agency, *Arms Control Impact Statements* (hereafter, ACIS) (Washington, D.C.: U.S. Government Printing Office, various years). The numbers cited do not include funding for research conducted by the Air Force or by the Advanced Research Projects Agency on directed energy and other technologies having potential relevance to BMD. Directed energy programs were being funded at about the same level as the Army BMD program at the end of the Carter administration. *FY82 ACIS*, p. 391. All these programs were gathered into the SDI budget starting in fiscal 1985.

31.   Don Oberdorfer, "A New Age of Uncertainty is Born," *Washington Post*, 4 Jan. 1985, p. 1; Robert Scheer, "Teller's Obsession Became Reality in 'Star Wars' Plan," *Los Angeles Times*, 10 July 1983, pp. VI-6 ff.

32.   David Hoffman and Lou Cannon, "President Overruled Advisers on Announcing Defense Plan," *Washington Post*, 26 Mar. 1983, p. 1; R. Jeffrey Smith, "Reagan Plans New ABM Effort," *Science* (4 Apr. 1983): 170-71; also Oberdorfer, and Scheer, note 31.

33.   Scheer, note 31.

34.   U.S., White House, *Report of the President's Commission on Strategic Forces* (Washington, D.C.: U.S. Government Printing Office, April 1983), pp. 7-8.

35.   Unclassified summaries of both reports are reprinted in Steven E. Miller and Stephen Van Evera, *The Star Wars Controversy, an International Security Reader* (Princeton, N.J.: Princeton University Press, 1986), pp. 273-317.

36.   See, for example, Edward Luttwak, "Why Arms Control Has Failed," *Commentary* (January 1978): 19-28: ("[it] has usurped the function of strategy and has become an end in itself . . . [used] to mask a renunciatory passivity in the face of the Soviet build-up. . . ."); Colin Gray, "Moscow is Cheating," *Foreign Policy* (Fall 1984): 141-52; and General Advisory

Committee on Arms Control and Disarmament, "A Quarter Century of Soviet Compliance Practices Under Arms Control Commitments: 1958–1983" (Washington, D.C.: October 1984, Mimeographed).

37. Thomas W. Graham, "Public Attitudes Towards Active Defense: ABM & Star Wars, 1945–1985," MIT Center for International Studies Report C/86-2 (May 1986). On Congress and the ABM Treaty, see, most recently, U.S. Congress, House, Committee on Armed Services, *National Defense Authorization Act for Fiscal Year 1987, Conference Report*, 99th Cong., 2d sess., H. Rept. 99-1001, 14 Oct. 1986, p. 30 (on ABM Treaty compliance) and pp. 449–50 (on refocusing SDI toward defense of strategic forces and command and control, and on developing near-term deployment options as a hedge against Soviet breakout from the Treaty).

38. At a June 1986 *Time* magazine symposium on SDI, Assistant Secretary of Defense Perle stated that the immediate goal of SDI is "the defense of America's capacity to retaliate." According to fellow participant Ambassador Nitze, "that is contrary to the White House view of the matter." *Time*, 23 June 1986, p. 18. In testimony before the U.S. House Appropriations Committee the previous April, Secretary Weinberger said that those favoring defense of missiles "don't understand the system and have not gotten the word." Cited by Lou Cannon, "Weinberger Warns Against SDI Trade-Off," *Washington Post*, 22 July 1986, p. A15. See also Walter Andrews, "Defending People Must Be SDI Goal, Says Weinberger," *Washington Times*, 3 Sept. 1986, p. 5; John Cushman, "Weinberger Gives Strategy Outline on Missile Shield," *NYT*, 13 Jan. 1987, p. 1; and George A. Keyworth, "Security and Stability: The Role for Strategic Defense," IGCC Policy Paper No. 1 (La Jolla, Calif.: Institute on Global Conflict and Cooperation, University of California, 1985), p. 7.

39. Caspar W. Weinberger, "Ethics and Public Policy: The Case of SDI," *Fletcher Forum* (Winter 1986): 1.

40. Scheer, *Los Angeles Times,* 10 July 1983. In a 1981 interview just after announcement of the Reagan Strategic Modernization Program, Weinberger said, "What we would like to have is something that makes ballistic missile attack on the U.S. fully ineffective." John Quirt, "Washington's New Push for Anti-Missiles," *Fortune* (19 October 1981): 142–48.

41. Fred C. Iklé, "The Vision Versus the Nightmare," *Washington Post*, 27 Mar. 1983, p. B8.

42. Fred C. Iklé, "Nuclear Strategy: Can There Be a Happy Ending?" *Foreign Affairs* (Spring 1985): 810–47. Robert Jervis observes that those who have the most nightmarish visions of the nuclear balance are those who "see the Russians as quite aggressive, highly motivated to expand, and willing to take significant risks to reach that end." Robert Jervis, *The Illogic of American Nuclear Strategy* (Ithaca, N.Y.: Cornell University Press, 1984), p. 122.

43. Reprinted in Miller and Van Evera, eds., *The Star Wars Controversy* note 35.

44. Paul Nitze, "Assuring Strategic Stability in an Era of Detente," *Foreign Affairs* (January 1976): 212–213, 226, 229.

45. Reprinted as "The SDI in U.S. Nuclear Strategy," *International Security* (Summer 1985): esp. 14–16, 19.

46. Warner Schilling, "U.S. Strategic Nuclear Concepts in the 1970s: The Search for Sufficiently Equivalent Countervailing Parity," *International Security* (Fall 1981): 48–79. See also U.S. Congress, Senate, Committee on Foreign Relations, *Hearings, U.S.-USSR Strategic Forces*, 93d Cong., 2d sess., 4 Mar. 1974 (testimony of James A. Schlesinger, pp. 9, 26); Walter Slocombe, "The Countervailing Strategy," *International Security* (Spring 1981): 18–27; Richard Halloran, "Pentagon Draws Up First Strategy for Fighting a Long Nuclear War," *NYT*, 30 May 1982, p. 1; and for the Weinberger clarification in his letter to the editor, see *Los Angeles Times*, 25 Aug. 1982.

47. Stephen M. Meyer, "Soviet Strategic Programs and the U.S. SDI," *Survival* (November/December 1985): 279–83.

48. James R. Schlesinger, "Rhetoric and Realities in the Star Wars Debate," *International Security* (Summer 1985): 5.

49. Thomas C. Schelling, "What Went Wrong with Arms Control?" *Foreign Affairs* (Winter 1985–86): 230.

50. Garthoff, *Détente and Confrontation*, note 26, pp. 771–73.

51. Michael MccGwire, "Dilemmas and Delusions of Deterrence," *World Policy Journal* (Summer 1984): 752.

52. See also Robert L. Arnett, "Soviet Attitudes Toward Nuclear War," in John Baylis and Gerald Segal, eds., *Soviet Strategy* (London: Croom Helm Press, 1981), pp. 60–61.

53. Stephen M. Meyer, "Soviet Nuclear Operations," in Ashton Carter, John Steinbruner, and Charles Zraket, eds., *Managing Nuclear Operations* (Washington, D.C.: Brookings Institution, 1987), pp. 471–72; also Benjamin S. Lambeth, "How to Think About Soviet Military Doctrine," in Baylis and Segal, note 52, p. 113.

54. Ibid., pp. 111, 114.

55. Meyer, "Soviet Nuclear Operations," note 53, pp. 528–29.

56. Fritz W. Ermarth, "Contrasts in American and Soviet Strategic Thought," in John Reichart and Steven Sturm, eds., *American Defense Policy*, 5th ed. (Baltimore: Johns Hopkins University Press, 1982), p. 66.

57. Stephen M. Meyer, "Soviet Theater Nuclear Forces, Part I: Development of Doctrine and Objectives," *Adelphi Paper No. 187* (London: IISS, 1984), pp. 27–29.

58. Meyer, "Soviet Strategic Programs," note 47, p. 285.

59. Lambeth, "How to Think," note 53, p. 110.

60. Stephen M. Meyer and Peter R. Almquist, *Insights from Mathematical Modeling in Soviet Mission Analysis (Part II)*, Report prepared for the U.S. Department of Defense Advanced Research Projects Agency (Cambridge, Mass.: Department of Political Science and Center for International Studies, MIT, 1985), pp. 25-26.

61. Robert Legvold, "Strategic 'Doctrine' and SALT: Soviet and American Views," in *Survival* (January/February 1979): 10-12.

62. See U.S., Congress, Joint Economic Committee, *Allocation of Resources in the Soviet Union and China, 1985*, 99th Cong., 2d sess. S. Hearing 99-252, pt. 11, 19 Mar. 1986, esp. pp. 9, 19, 95, 119.

2 THE ABM TREATY

Arms control agreements age in dog years. The naval arms limitation agreements of the 1920s, cast in a technologically simpler but politically and economically more volatile age than our own, were losing their effectiveness only 10 years after their signature; within 15 years they were dead, victims of advancing technology and national ambitions.[1] After fifteen years, the Anti-Ballistic Missile Treaty is very much alive, although likewise under pressure. Some of its problems are technology related and some are political. These are reviewed in detail below, along with an analysis of the costs and benefits of the Treaty. However, some of the Treaty's problems derive from the nature of arms control itself and the verification efforts it requires; these issues are treated first.

THE BURDENS OF ARMS CONTROL

Bilateral arms control agreements are made between political/military adversaries. There must be some common interests or no treaty will ensue, but to the signatories, the primary benefits derive from constraints placed on the adversary's forces (which constitute the threat). Limits on one's own forces by and large are viewed as a cost and not a benefit of agreement, certainly by the affected armed services and usually by their political leadership as well. This explains why past arms accords have tended to rechannel arms production into other, unconstrained systems (cruisers in the 1920s, cruise missiles in the 1970s): agreements have aimed at particular classes of threat, not arms in general, much less at fundamental political differences, and those remain powerful force drivers. Although the United States and the Soviet Union did pursue limited political accommodation in parallel with SALT, their respective definitions of "détente" were incompatible.[2]

The achievement of threat control—constraint on the other side's forces—is complicated by the requirement of symmetry in the ultimate agreement. Even an apparent lack of symmetry can cause problems for an agreement. In the early 1970s, asymmetric limits in the SALT I Interim Agreement led the U.S. Congress to mandate equal arms ceilings in future U.S.-Soviet accords.

Standards of Verification

The focus on threat control, plus simple self-interest, account for the heavy emphasis on verification of the other side's compliance with the terms of arms control agreements. Since verification is both a political and judgmental process, standards of compliance change with domestic political climates, with decisionmakers' predispositions, and with shifts in bureaucratic interests and power, so compliance judgments may change from one administration to the next without any changes in the underlying base of intelligence data. During the Carter administration, the standard was "adequate" verification; that is, a capability sufficient to permit the United States to respond to Soviet violations quickly and strongly enough to offset any gains that it might seek with such behavior. With regard to the ABM Treaty, such responses would include rapid deployment of penetration aids on offensive missiles. Under President Reagan, the standard has become "effective" verification, a term not clearly defined but implying a more rigorous approach.

How any given administration approaches verification reflects, in particular, the internal balance between those who believe that the Soviet Union is "a committed treaty violator" and those who believe that it is just "an extraordinarily difficult negotiating partner."[3] The "committed violator" thesis was set forth most explicitly in an October 1984 report of President Reagan's General Advisory Committee on Arms Control and Disarmament (a report not formally reviewed or endorsed by the Reagan administration). In its view, because Soviet interests and intentions are so antithetical to U.S. interests, the only satisfactory standard of verification is "absolute," that is, certification of Soviet *non*-violation. Proving a negative in this situation is sufficiently difficult that Soviet compliance cannot be assured, thus verification is never good enough and arms control is inherently undesirable.[4] Between this extreme and one that accepts Moscow's word alone as sufficient evidence of compliance lies a range of possibilities, a range that entails concepts of risk management.

Uncertainty and Risk

Uncertainty in compliance monitoring and the potential risk to national interests entailed by an arms control agreement may be thought to move in lockstep, but they do not always do so. Uncertainty in verification is "largely an objective measure" of a monitoring system's "ability to provide data of unambiguous meaning." Risk, on the other hand, is an "inherently subjective" measure of the consequences or costs likely to follow from some decision or course of action.[5] Indeed, the two may vary inversely, with important and, as Meyer notes, counterintuitive implications for the design and implementation of arms control agreements, and for the evaluation of compliance problems, solutions to them, and measures to strengthen the agreements in question.[6] Some examples illustrate.

The verification process can return any of a range of judgments about a given activity at any given time, from "apparently compliant," to "anomalous, bears watching," to "questionable, needs resolution," to "inadvertent violation," to "deliberate violation."[7] The *uncertainty* reflected in those judgments is lowest at either end of the range and higher toward the middle. However, the implied *risk* to U.S. interests increases from one end of the range to the other.

Meyer uses as an example the uncertainties inherent in monitoring Soviet tests of air defense systems in an ABM mode, which are forbidden under the Treaty. Any such testing, he observes, must be done clandestinely and thus could pose detection problems for U.S. intelligence:

> Compare this situation to one in which the ABM Treaty no longer exists. The Soviets would be free to test and upgrade their vast air defense network for ABM missions, and it would be absolutely clear to U.S. monitoring systems that that was what they were doing. In many respects, the first situation poses far fewer risks to U.S. national security than does the second, even though there may be greater uncertainty in the first case.[8]

More generally, risks posed by *uncertainties in monitoring* vary with the military significance of undetected violations, the military significance of activities permitted under an agreement that could undermine its viability, the net security benefits derived from a properly observed agreement, and the options available to offset the consequences of cheating by the other side.[9] Risks posed by *detected violations* must be similarly weighed according to the seriousness of the violation, its most likely source (that is, accident, technical foulup, or deliberate), how the Soviet Union responds when queried about it, and what options are available to the United States if the Soviet response is unsatisfactory. The chance of *false detections* will rise if the system is pressed not to miss any potential violations, the risk being that compliance challenges based on false detections will undermine the viability of an agreement.[10]

One approach to reducing uncertainty in monitoring would be to emphasize total bans over partial limits. For example, a global ban on mobile ballistic missiles and launchers of type "X" would require the verified dismantling of all existing missiles and launchers and of their production and support facilities. Any subsequently detected missiles, launchers, or facilities would be a clear violation. On the other hand, an agreement that entailed partial limits would leave an agreed-upon fraction of missiles and launchers in the field, requiring additional inspection of specified deployment areas, constant accounting of the number and location of remaining systems, on-site monitoring of their production and maintenance facilities, and agreement on non-encryption of test-flight telemetry.[11]

Risk may be reduced and U.S. interests advanced, however, even with limits that *do* involve significant uncertainties in monitoring. Limits set below the minimum detection threshold of current monitoring systems may still be desirable if they "minimize the military potential" of the constrained system or technology.[12] In effect, they would be akin to the bans discussed above: Any activity that *is* detected in the prohibited system or activity constitutes a *prima facie* violation. The more generous the buffer between permitted performance levels and a real weapons capability, the lower the risk, despite high uncertainty. Consider, for example, the case of "directed energy" weapons. The power requirements for laser communications are on the order of "tens of watts," which corresponds to a beam "brightness" or intensity of 10^5 to 10^{14} watts/steradian, whereas the brightness required to destroy a nominal ICBM booster at long range may be on the order of 10^{20} watts/steradian, or at least a million times greater.[13] Suppose that U.S. national technical means of verification (NTM) could reliably detect laser tests down to brightnesses of 10^{18} watts/steradian.[14] If the goal were to limit laser weapon development but not the use of lasers for communications, an agreement limiting laser brightness to 10^{15} watts/steradian might be more desirable, from the standpoint of limiting weapon potential, than a limit of 10^{18}. Technological advances in monitoring would eventually lower the early detection threshold. Meanwhile, laser development and testing could not legally come anywhere close to weapon capability; there would be legal grounds for challenging suspicious activities at much lower than weapons-grade power levels; and existing NTMs could detect a real weapons-grade test.

Arms control measures can be evaluated, then, according to how much they reduce risk, relative to an unconstrained case, and how much uncertainty they entail at any given level of risk. Before evaluating current shortcomings in the ABM Treaty according to this rubric, it is important to have a basic sense of the Treaty's costs and benefits to U.S. interests.

BENEFITS AND COSTS

Rationales for U.S. acceptance and continued adherence to the 1972 ABM Treaty generally fall into the following categories: the cost of armaments, impact on the arms competition (including setting the stage for limits on offensive arms), impact on crisis stability, effects on strategic uncertainty, and cohesion of the NATO alliance.

Arms Cost

Many analysts agree that the ABM Treaty allowed both sides to avoid deploying costly but ineffective ABM systems, and offensive responses to them, during the 1970s. Soviet actions absent the ABM Treaty are unknowable, but U.S. intelligence in 1969 projected Soviet ABM deployments for the mid-1970s to be between 5,000 and 9,000 interceptors, and in 1972 estimated that 8,000 to 10,000 Soviet ABM interceptors could be deployed by 1980 absent agreed-upon limits.[15]

Other analysts believe that the Treaty largely ratified the two sides' prior decisions to postpone deployments, citing the 1968 halt in Soviet construction of the Moscow-based ABM system and U.S. Senate refusal to fund area defense, and near-refusal to fund anything, in 1969–1970. George Rathjens suggests that only because of its role as a bargaining chip in the SALT negotiations was Safeguard deployed at all.[16]

While there is some disagreement as to whether the Treaty obviated large-scale expenditures in the 1970s, it is hard to contest the proposition that if the Treaty remains in force it will allow both sides to avoid future, competitive expenditures on strategic BMD that could easily exceed $500 billion for development, procurement, and the first ten years of operation.[17]

Arms Competition

The United States undertook development of multiple independently targeted reentry vehicle (MIRV) technology primarily to counter Soviet ABM deployments then underway or projected.[18] In his September 1967 speech announcing the decision to deploy Sentinel, Defense Secretary McNamara stressed the dangers of the "action/reaction" phenomenon.[19] In statements before the Supreme Soviet in 1972, Soviet military leaders stressed the restraining effects of the ABM Treaty on offense-defense interaction. Chief of the General Staff Kulikov and Defense Minister Grechko both endorsed the Treaty as "preventing the emergence of a chain reaction of competition between offensive and defensive arms."[20] And in June 1972 hearings on the ratification of the ABM Treaty before the Senate Foreign Relations Committee, National Security Advisor Henry Kissinger stated that, although SALT had failed to produce a MIRV ban, "by setting a limit to ABM defenses, the treaty . . . reduces the incentive for continuing deployment of offensive systems."[21]

In the 1980s, Treaty critics could refer to these statements, point to the rapid growth in numbers of Soviet strategic warheads in the 1970s (due to MIRVing of Soviet missiles within SALT launcher limits), and declare the ABM Treaty to be a failure.[22]

Yet penetration of Soviet missile defenses was only one of many reasons for deployment of MIRV. For the U.S. military, MIRVs promised to increase strategic target coverage much more cheaply than would increasing the number of launch vehicles, while also providing for more accurate delivery of weapons against hardened targets. For the Soviet military, MIRVs offered a means to efficiently exploit the great throwweight of their ICBMs for counterforce targeting. Unlike late-1960s ABM technology, MIRVs seemed sure to work.

This is not to say that the Treaty abetted the deployment of MIRVs. Constraints on BMD permit a given number of warheads to accomplish missions that would need many more warheads if significant attrition from active defenses were expected. Depending on the "leverage" enjoyed by the defense (a function of the fraction of incoming warheads that it must engage to accomplish its mission), each interceptor deployed will prompt a conservative force planner to assign one or more additional warheads to attack the defended target set.

The Treaty can be viewed, then, as having set necessary but not sufficient conditions for offensive strategic arms control measures. By holding defenses fixed, it laid the groundwork for offensive limits in SALT I and II. That a better structure was not built later does not reflect upon the quality of the original foundation.

Crisis Stability

Western deterrence theory places great emphasis on structuring forces so as not to give one's opponent (or oneself) an incentive to strike first. Vulnerable offensive forces theoretically provide some incentive in that direction. Adding moderately capable area defense is thought to increase the theoretical incentive further: A first strike by country A penetrates country B's defenses and thins out B's partially or wholly vulnerable offense; B's offense, although in better shape than it would have been if completely undefended, manages to respond with only a ragged volley. A's full-strength defenses thin out that volley, with the intent of limiting damage in A to acceptable levels. The theoretical incentive structure varies somewhat depending on the relative sizes of the defenses.

However, the incentive structure is fundamentally affected by the basic vulnerability of offense and defense. Largely survivable offensive forces and associated command, control, and communications (C^3) could hand an attacker retaliatory damage equivalent to that wrought by a first strike. If both sides had a survivable offense, and if that survivability were largely independent of the efficacy or vulnerability of BMD, the probability that either side would view a first strike as being preferable to other alternatives in a crisis would be low, even if both sides deployed partial area defenses. Nor would pressures to launch on tactical warning be particularly acute. Should war break out, the respective levels of damage incurred would vary only with the effectiveness of respective defenses in dealing with countervalue attack. Since conservative planning would tend to undervalue one's own defenses and overvalue the other side's (unless disparities in defensive capabilities were clearly and extremely favorable), there would be little rational incentive to strike first. However, the disparities would be made favorable and incentives could be altered radically if either side's offense were partially vulnerable, depended on its defense to alleviate that vulnerability, and if its defense were also vulnerable to defense suppression attacks.[23]

The primacy of offense survivability in rational models of crisis stability and the potential vulnerability of defensive systems suggest reliance on other options (mobility, concealment, hardening) to improve offense survivability. Thus, in ruling out force defenses, the Treaty does not *prevent* steps to enhance crisis stability and, in ruling out area defense, it may *contribute* to crisis stability by reducing undesirable sorts of strategic uncertainty and maintaining desirable ones.

Strategic Uncertainty

The ABM Treaty rendered the strategic planning environment more predictable, by holding constant some very messy variables that would otherwise complicate strategic force planning and targeting equations. This is arms control in the managerial sense, the sense in which the enterprise appeals to military planners. Since the 1972 Treaty, it has been easier to calculate warhead requirements for any given targeting strategy. Expected attrition due to active defense is only a nuisance factor for U.S. planners (because of the Moscow ABM system), only a slightly more significant factor for French and British planners and, at least since 1976 (when the one U.S. Safeguard site was dismantled), not a factor for Soviet planners. In effect, the United States traded the future risk of fixed-silo ICBM vulnerability for improved prospects of successful retaliation by all U.S. strategic forces surviving a Soviet first strike: ICBMs, submarines, and bombers.[24]

Proponents of the window of vulnerability theory argue that the Treaty made the Soviet targeting task too easy, because it guaranteed that U.S. ICBM vulnerability would simply be a function of increased Soviet ICBM accuracy and reliability.[25] They conclude that Soviet uncertainty was reduced too much and needs to be increased. According to the January 1985 White House statement on SDI, effective BMD would reinforce deterrence, first and foremost by "significantly increas[ing] an aggressor's uncertainties" regarding prospects of success in attacking the United States and "would restore the condition that attacking could never leave him better off;" that is, effective BMD would restore crisis stability, the implication being that current Soviet incentives to strike first in a crisis are unacceptably high.[26] The Scowcroft Report notwithstanding, a chill wind still blows through the window of vulnerability.

However, depending on circumstances, deployment of BMD could both create and dispel uncertainty. Sometimes its effect would favor Western interests and sometimes not, as the following assessment suggests.

With respect to the adequacy of deterrence under current circumstances, it is important to consider whether a disarming counterforce attack by the Soviet Union on the United States would be a straightforward and low-cost task, so straightforward that deterrence requires an assist. It is generally agreed that even if a high percentage of U.S. ICBMs were destroyed in a Soviet attack, no less than 30 percent of the bomber force and one-half to two-thirds of the missile submarine (SSBN) force would survive to retaliate. By the mid-1990s, (the earliest feasible time for a BMD system based on non-nuclear intercept technology to reach initial operational capability), the surviving submarine force alone would include more than 1,700 accurate counter-force warheads on D5 (Trident II) missiles—greater than the total force of high-accuracy warheads to be deployed then on U.S. MX and Minuteman III ICBMs. Could rational Soviet decisionmakers afford to act on the prospect of missing roughly 75 percent of U.S. retaliatory forces, including one-half of the U.S. prompt retaliatory counterforce capability, on a gamble that U.S. strategic C^3 could be totally disabled or—an even greater gamble—that U.S. leaders would not shoot back?[27] If the Soviet Union lacked significant territorial missile defense, they clearly could not. However, the availability of such defenses could make a decision to preempt-and-defend easier to reach in a crisis. The first question to ask about a world with such defenses, then, is whether *current* uncertainties about the prospects for a damage-limiting first strike would be sustained in the minds of Soviet leaders.

Second, if the principal concern is maintaining crisis stability, one would want to be quite selective about what sorts of uncertainty were promoted. It is not correct to assume that adding uncertainty to Soviet calculations is always beneficial to Western interests. In-deed, for some purposes, near-certainty in Soviet perceptions is desirable: For example, near-certainty that the United States *will* retaliate if attacked, or near-certainty that the United States will *not* preempt in a crisis. Defenses could contribute two sorts of crisis un-certainty, only one of which is addressed in Reagan administration policy (the sort that diminishes a would-be attacker's confidence of a successful attack). The second and less desirable contribution of defense to uncertainty would affect one side's ability to assess the other side's intentions; if defenses were thought to encourage a preempt-and-defend strategy by one side, it could increase the pres-sures felt by the other side to preempt. Defenses would lower the efficiency and perhaps also the effectiveness of a defensive-preemp-tive attack, but would *not decrease its probability* if preemption seemed, in a deep crisis, the best choice in a bad lot.

As Charles Zraket observes, "the ability to survive and to fight, im-plied by BMD, might well lower the nuclear threshold rather than strengthen deterrence. . . .;" moreover, "the range of uncertainties introduced by BMD may in fact result in *incalculable outcomes* [his emphasis]."[28] In short, uncertainty cuts more than one way, and BMD may have more than one effect on uncertainty. The Treaty re-duces certain elements of uncertainty in a very complex strategic environment. It does not in any sense make that environment a simple one.[29]

Alliance Cohesion

The linchpin of the Western Alliance, notes Christoph Bertram, "has been the ability of U.S. strategic nuclear forces to deter a possible Soviet attack," and Soviet vulnerability to nuclear strikes has been a prerequisite to the credibility of that deterrent threat. U.S. vulnerability plays a role as well: "Europeans are profoundly convinced that their security rests on America's recognition of its *own* vulnerability. For Europeans, American-European solidarity is not just a matter of declared interests, but of shared fate."[30] The ABM Treaty, in this view, ensures that the United States shares Europe's vulnerability; that their fate in any future NATO-Warsaw Pact war would indeed be a common one.

On the other side of the coin, one finds recurring disputes about the credibility of NATO's nuclear deterrent threats as a balance to the conventional forces available to the Soviet Union and its Warsaw Pact allies.[31] Fueling this concern is the growth of the Soviet nuclear threat to U.S. territory and the potential impact of this on U.S. willingness to defend Western Europe. If, due to its own vulnerability, the United States looked unlikely to make the ultimate sacrifice on behalf of its allies—"to trade New York for Bonn" in the event of war—the credibility of NATO's deterrent posture would be undermined and with it Western Europe's security and long-term prospects for peace and political independence. Theoretically then, the prospect of greatly reduced U.S. vulnerability to Soviet missile attack ought to be welcomed in Europe as a contribution to deterrence. However, not everyone sees it that way.

Ivo Daalder notes the European NATO allies' "interpretation of flexible response . . . as a seamless web, where the destruction of just one part risks the detonation of the whole machine." In starting even a conventional conflict, the Soviet Union would be risking "automatic" escalation to the level of a strategic nuclear war. Uncertainty about the course of the war and the risks it would entail thus are an integral part of NATO's basic deterrent. Defense of the U.S. homeland, Daalder suggests, "detaches the American strategic forces from this interconnected deterrent whole. Once again, the potential use of these forces would become subject to an American decision and could thus be withheld. A war confined to Europe would hence-forth again become a possibility."[32] Agrees Bertram, SDI "indi-cates that the United States wants to escape from the risk of a Euro-pean conflict. . . . [T] here is no concealing the fact that the United States is now actively pursuing a course that, if implemented, will amount to restricting conflict to Europe. This policy profoundly weakens both deterrence and Alliance cohesion."[33]

On the other hand, Jacquelyn Davis and Robert Pfaltzgraff, who strongly favor BMD, voice concerns about the implications for NATO of effective *Soviet* BMD:

> Clearly, the purpose of a [Soviet] strategic force posture based on a mix of offensive and defensive means . . . is to deprive the United States of the abil-ity to pose a credible nuclear threat to the Soviet Union. . . . It must be pru-dently assumed that one of the principal objectives of such a Soviet force posture will be the decoupling of the United States from NATO-European security. . . .[34]

Davis and Pfaltzgraff also emphasize the danger to NATO of *current* Soviet strategic missile defense capabilities. NATO, in their view, will soon find itself looking across a BMD gap unless swift action is taken to close it.[35]

However, they do not extend their argument to the case in which both sides have very good BMD. Sidney Drell and others conclude that if both superpowers deployed very good defenses, then "the Soviet Union would presumably then find it easier to reach a deci-sion to attack NATO Europe conventionally since it could protect itself against nuclear attack and ignore the U.S. nuclear guarantee to Europe."[36] That is, joint deployment of defenses could also be de-coupling, and thus destabilizing.

In this context, it might be noted that Soviet incentives with re-spect to war initiation, insofar as they are affected by U.S. nuclear threats against the Soviet homeland, *would be no different* in the bilaterally defended scenario than in the unilateral scenario that Davis and Pfaltzgraff fear. Under the defense conditions posited (mutual capability to limit damage against respective first strikes), the Soviet Union would expect to suffer equivalent damage levels whether or not the United States deployed its own missile shield. Deterrence would devolve upon battlefield forces, both nuclear and conventional, bringing us back full circle to European concerns about the confinement of war and its consequences to Europe alone.

However, strategic BMD is unlikely to work so well that both superpowers will ever render themselves totally impervious to nuclear attack, at least given currently deployed levels of offense. Debates about deterrence and coupling within NATO thus would continue after defense deployments, since at base these are political debates about the reliability of U.S. assurances (for which the various arguments over hardware and doctrine are essentially proxies). What might be the effect of partial defenses (U.S. and Soviet) on perceptions of U.S. reliability? If zero BMD plus U.S. forces in Western Europe make for shared vulnerability but at the risk that Washington could be intimidated in wartime, and if very good defenses would decrease the intimidation factor but leave Washington indifferent to Europe's plight, then the impact of partial defenses might be expected to fall somewhere inbetween: increasing U.S. willingness to strike the Soviet Union if required by NATO, without entirely removing the incumbent risk of a European war to either superpower's homeland. The downside of partial defenses, however, may be decreased crisis stability, for reasons discussed previously, and thus increased chances of war.

The Treaty also has a separate, political importance for NATO. As Daalder and Lynn Whittaker note, because of the Treaty's image in Europe as the centerpiece of East-West arms control and détente, any U.S. move to withdraw from it could have severe negative repercussions for U.S. leadership of NATO. Moreover, any renegotiation of the Treaty (or any unilateral Soviet action) that resulted in further deployment of BMD by the Soviet Union would have a more immediate and telling impact on Allied nuclear forces than on those of the United States. In short, either unilateral or bilateral actions that loosened constraints on BMD would be detrimental to NATO cohesion.[37]

CONTENT AND PROBLEMS

At this point, the discussion turns from general principles, costs, and benefits to the detailed structure of the ABM Treaty itself and the issues that have arisen—as a function of the passage of time, advances in technology, and changes in the political climate—to pose challenges for both the short- and long-term viability of the agreement. The issues discussion first addresses hardware that is currently deployed or believed to be deployable; then it addresses issues involving development and testing of new BMD (and near-BMD) technologies, including analysis of the Reagan administration's reinterpretation of the Treaty.

The Terms of Agreement

The Treaty is an agreement of unlimited duration in which each party, as stated in Article I, undertakes "not to deploy ABM systems for a defense of the territory of its country and not to provide a base for such a defense, and not to deploy ABM systems for defense of an individual region except as provided in Article III." Article II defines an ABM system as "a system to counter strategic ballistic missiles or their elements in flight trajectory, currently consisting of: ABM interceptor missiles . . . ABM launchers [and] . . . ABM radars," which are those "constructed and deployed for an ABM role, or of a type tested in an ABM mode." The latter phrase is not defined in the Treaty itself, but was the subject of a unilateral U.S. statement during negotiations and two classified Agreed Statements subsequently worked out in the SALT Standing Consultative Commission (SCC). The prohibition on "testing in an ABM mode" is the principal tool by which non-ABM systems are prevented from attaining prohibited ABM capability.[38]

Article III reiterates the ban on ABM deployment, excepting one deployment area around the national capitals and one around an ICBM launcher deployment area. A 1974 protocol to the Treaty later reduced the permitted deployment area to just one apiece. Within that deployment area there may be 100 ABM launchers and "no more than 100 ABM interceptor missiles at launch sites." Radars within the deployment areas are also limited. Radars defending the national capital are confined to 6 small subareas but are not otherwise limited in quantity or quality; sites for defending ICBMs are limited to 2 "large phased array radars" (LPARs) and 18 "smaller" radars. An Agreed Statement (D) issued in connection with Article III prohibits the deployment of ABM systems or components "based on other physical principles" that are "capable of substituting" for traditional components. The implications of this Agreed Statement for restraints on such "exotic" components have been the subject of much recent dispute.

Article IV allows deployment of ABM components, including up to fifteen ABM launchers at "current or additionally agreed" test ranges. The definition of Soviet test ranges was not completely worked out until 1975.

Article V prohibits the development, testing, or deployment of ABM systems or components that are sea-, air-, space- or mobile land-based. It also prohibits development, testing, or deployment of rapid-fire and rapid-reload launchers. An associated Agreed Statement (E) prohibits the development, testing, and deployment of ABM interceptors that have more than one independently guided warhead, and a Common Understanding (C) reflects U.S.-Soviet agreement that Article V rules out deployment of ABM launchers and ABM radars that are not "permanent, fixed types."

Article VI blocks circumvention of the Treaty by means of non-ABM technologies. The parties undertake not to give non-ABM missiles, launchers, or radars the "capabilities to counter" strategic ballistic missiles or their elements and further agree not to test such equipment in an ABM mode. The parties also agree "not to deploy radars for early warning of strategic ballistic missile attack" in the future except along the periphery of their national territory, oriented outward. An Agreed Statement (F) creates an important exception to this constraint for radars used "for the purposes of tracking objects in outer space or for use as national technical means of verification."

Article VII permits modernization and replacement of permitted ABM systems and components. It legitimates, for example, current upgrade work on the Moscow ABM system. It does not address early warning radars or other non-ABM systems or components.

Article IX, a further measure to prevent circumvention, forbids the transfer of ABM systems or components to third parties, or their deployment outside either signatory's national territory. An Agreed Statement (G) further bars the transfer to third parties of "technical descriptions or blueprints" of ABM systems or components.

Article XII provides that monitoring of compliance shall be by national technical means (NTM). It prohibits interference with NTM and use of deliberate concealment measures to impede NTM monitoring. This Article was the first explicit acknowledgment by the Soviet Union of another state's right to conduct limited information-gathering operations with respect to Soviet territory, and it was the first Soviet public pledge not to interfere with such operations.[39]

The only other bilateral agreements with similar language (the SALT I Interim Agreement and the SALT II Treaty) are not now in force.

Article XIII outlines the responsibilities of the SCC, which is a unique bilateral forum charged with considering questions of Treaty compliance, unintended interference with NTM, procedures and dates for dismantling and destroying hardware as required by the Treaty, changes in the strategic situation, amendments, and "proposals for further measures aimed at limiting strategic arms."

Article XIV provides for Treaty amendment (which may be proposed by either party at any time), and for reviews of the Treaty every five years.

Article XV sets the duration of the agreement (unlimited) and the conditions for withdrawal from it (upon six-months' notice, should "extraordinary events related to the subject matter of this Treaty have jeopardized [a party's] supreme interests"). Determination of conditions for withdrawal is wholly unilateral, although the 1969 Vienna Convention on the Law of Treaties (which the United States supports in principle but has not ratified) provides generally accepted guidelines for withdrawal and for dealing with a "material breach" of Treaty obligations: "A material breach of a bilateral treaty by one of the parties entitles the other to invoke the breach as a ground for terminating the treaty or suspending its operation in whole or in part." The Vienna Convention defines a material breach as "(a) a repudiation of the treaty not sanctioned by the present Convention; or (b) the violation of a provision essential to the accomplishment of the object or purpose of the treaty." As will be seen, Reagan administration charges that the Soviet Union may be laying a "base" for a missile defense of its national territory come close to stating that a material breach of the Treaty is underway.[40]

Deployment-Related Issues

Principal issues here relate to construction of large phased-array radars; rapidly deployable ABM-associated radars and reloadable launchers; and problems related to air defense upgrades.

Large Phased-Array Radars. The heart of the ABM Treaty, Article I, forbids both the deployment of an ABM system for the defense of national territory and the deployment of a "base" to support such a system. The latter prohibition is buttressed by Common Understanding (A), which sets a minimum separation of 1,300 kilometers between the two originally permitted ABM sites; and by Article VI, which bans upgrades and testing in an ABM mode and which rules out laying a base for territorial missile defense via air defense upgrade.

The prohibition of a "base" builds lead time into any effort to deploy prohibited defenses. Thus, it gives a party observing prohibited activities greater time and opportunity to contest and to respond to them. What constitutes a "base" for defense of territory has never been defined bilaterally, but in the SALT I negotiations the United States pressed for limits on LPARs on the grounds that unlimited LPAR deployment could provide such a base. Because LPARs take many years to build, they would be the longest lead-time items in any ABM system; thus, restraining them would build the biggest possible deployment buffer into the Treaty. The Soviet Union resisted U.S. efforts to include any radar limits in the Treaty until late in the negotiations.[41]

Article VI limits any future deployment of early-warning LPARs to the periphery of national territory, facing outward. Agreed Statement (F) constrains the deployment of all LPARs with a power-aperture product (mean emitted power in watts multiplied by the antenna area in square meters) greater than three million, except as provided in Articles III (at ABM deployment areas), IV (ABM test ranges), and VI (on the periphery, as noted). However, it goes on to except deployment of LPARs "for the purpose of tracking objects in space or for use as national technical means of verification" without regard to power-aperture product, location, or orientation. That language protected U.S. options to upgrade its global spacetrack network and to deploy LPARs as necessary to monitor Soviet missile tests. The large Cobra Dane LPAR on Shemya Island at the tip of the Aleutian chain operates as such an NTM radar. Assembled from the remnants of a Missile Site Radar built for the pre-Treaty Safeguard ABM site at Malmstrom Air Force Base, Cobra Dane, because of its location and orientation, also serves legitimately as an early-warning radar. The ship-based Cobra Judy radar also functions as an NTM; without the NTM exception, Cobra Judy would be a Treaty violation, a sea-based radar tested repeatedly "in an ABM mode" while watching Soviet strategic reentry vehicles (RV). The Soviet Union has claimed a spacetrack exemption for one of its LPARs, a claim the United States rejects.

Both sides are completing construction of a ring of early-warning LPARs around their respective national territories. The United States has five operational LPARs: the former Safeguard Perimeter Acquisition Radar in North Dakota; two Pave Paws radars, on Cape Cod and in California, for early warning of SLBM attack; an older phased-array spacetrack radar in Florida that doubles as an SLBM warning radar; and Cobra Dane. Four more LPARs are under construction: two Pave Paws radars, in Georgia and Texas; one Pave Paws-type radar at the Thule, Greenland site of the U.S. Ballistic Missile Early Warning System (BMEWS); and one at the BMEWS site in Fylingdales, England. (There is a third BMEWS site, on U.S. territory in Clear, Alaska, not currently slated to receive an LPAR.)[42]

The Soviet Union currently has 8 LPARs deployed: 3 at the Moscow ABM system, and 5 operational Pechora-class early-warning radars (named for the location of the first unit). Four more Pechoras are under construction. Three of these are on the western periphery of the Soviet Union; one is in western Siberia, near Krasnoyarsk. All save the Krasnoyarsk radar comply with the Treaty, but the three western radars have been tagged by the U.S. Defense Department as being potential building blocks for the "base" of a nationwide Soviet ABM system.[43] Other building blocks are said to be "transportable" radars and reloadable ABM launchers.

Krasnoyarsk. Although this radar will not become operational until the late 1980s, U.S.-published sketches of the two radar buildings show structures very similar to other Pechora-class radars, but having a somewhat smaller transmitter face.[44] The installation's 120-degree radar fan will be oriented to the northeast, filling a gap in the Soviet early-warning net for missile attacks launched from the northern Pacific.[45] According to Article VI, an early-warning radar with this orientation should have been located on the Pacific coast or in the outer, Arctic reaches of Siberia. Terrain, climate, and sheer inaccessibility may have ruled out the latter location (one report suggests that the Soviet Union had already experienced failure in an effort to maintain radar stations on the Siberian periphery[46]). A radar sited in northeastern Siberia would have provided better cross-range tracking data on U.S. ICBMs bound for the western Soviet Union than will Krasnoyarsk, but it would not have fully closed the coverage gap. Krasnoyarsk will do a better job of tracking long-range SLBM warheads launched from the mid- to western Pacific and targeted on points from Lake Baikal westward. Warheads so targeted would reach apogee roughly as they passed over the Soviet coast and entered within the radar's range and line of sight. Warheads aimed at targets east of Baikal (that is, at most of the Soviet far east) would tend to underfly the radar.

Krasnoyarsk's technical performance can be extrapolated from Pechora-class radars already operational. Although some U.S. Defense Department officials have labelled Krasnoyarsk an "ABM radar," its likely operating frequency, at 150 megahertz (MHz) in the very high frequency or VHF range, will likely give it little capacity to discriminate RVs from decoys and booster fragments, or to guide an ABM interceptor. Indeed, the United States gave up VHF frequencies for ABM search and acquisition radars at about the same time the Nike-Zeus project was abandoned, in favor of higher frequencies capable of better resolution and better resistance to disruption by nuclear effects. The Safeguard Missile Site Radar, for example, operated at 3,600 MHz (in the super high or SHF band), and U.S. Pave Paws and BMEWS radars operate around 400 MHz (ultra high frequency or UHF).[47]

The Krasnoyarsk radar appears to be a product of political/operational tradeoffs, an example of military necessity seeming to outweigh an inconvenient Treaty obligation. The decision to build the system would have been made in the late 1970s. One interpretation is that it was a deliberately planned Treaty breach only five years after signature. Another interpretation, consistent with Soviet compliance patterns generally, is that the letter of the Treaty left a loophole; a radar coverage gap needed plugging; a decision was made to build the radar and see if Washington complained; if complaints were made, efforts would be made to preserve the radar; and other options would be considered as necessary. A third interpretation views Krasnoyarsk as being in part a riposte to certain U.S. radar installations about which Moscow voiced objections. (The Soviet Union challenged U.S. construction of the Cobra Dane radar in the SCC in 1975, and similarly objected to Pave Paws construction in 1978.)[48]

Thule and Fylingdales. In response to the U.S. challenge on Krasnoyarsk, Moscow challenged the Thule/Fylingdales upgrades, and later offered to stop construction at Krasnoyarsk if the United States would halt construction at both its sites.[49] The U.S. position is that early-warning radar installations already in existence when the Treaty was signed may be modernized regardless of location. The Thule radar is a direct replacement for an older, mechanically-scanned tracking radar damaged by fire. It will replicate the 240-degree coverage of the original BMEWS radars. The Fylingdales radar is also "grandfathered" (exempt from limitation), it is argued, although its construction site will be several miles away from the original radars and its azimuthal coverage will increase from 180 to 360 degrees.

Once again, not everyone agrees. Critics talk about "conversion" at Thule and Fylingdales, not modernization. They note that, while Article VII provides for modernization of ABM systems and components, early-warning radars are not ABM radars; they are singled out in the Treaty for separate treatment. Indeed, if they were ABM radars, they could be deployed legally only at the two permitted ABM sites (Grand Forks and Moscow). Critics also cite internal U.S. government reports on the issue. First, an October 1971 report for the Verification Panel (Kissinger's policymaking group for SALT) reportedly concluded that the Treaty would allow only the Clear, Alaska BMEWS site to be upgraded to phased array. Since the United States then had no intention of upgrading BMEWS to phased arrays (perhaps because early-warning satellites by that time were a proven technology), it went along with the provision. Second, a 1984 National Security Council report on the Krasnoyarsk radar reportedly concluded that since Krasnoyarsk was an LPAR and did not specifically fall under one of the Treaty's enumerated exceptions, it was prohibited. By extension, it is argued, the Thule and Fylingdales LPARs, as early-warning radars not on the periphery of U.S. national territory, would be prohibited as well.[50]

The Soviet Union has claimed a site-location exemption for this radar because it is to be used for spacetracking, a purpose exempted under Agreed Statement F. Although the Statement gives no guidance with respect to distinguishing an early-warning radar from exempted radars, Treaty constraints on ABM radars clearly would be vitiated if either party were free to build LPARs at will so long as they painted "NTM" or "spacetrack" in large letters across the top. Indeed, some contend that even if current LPAR constraints are followed to the letter and the exemptions clause is not abused, a substantial fraction of the radar base for nationwide missile defense can still be built.

Baranovichi. The latest (and perhaps final) three radars of the Pechora-class early-warning network are newly under construction in the western Soviet Union (at Skrunda, in Lithuania; Baranovichi, in Belorussia; and Mukachevo, in the Ukraine). They are likely to be fully operational in the early- to mid-1990s. Baranovichi, unlike the other two sites, is not co-located with an older-type early-warning radar. Defense Department officials note that the resulting redundancy in radar coverage (Baranovichi is about midway between the other two installations) is useful for BMD tracking and battle management.[51]

Several observations are in order here. First, U.S. Pave Paws coverage will have a better than 50 percent overlap in places, when the final two southern sites are operational at the end of the decade. Second, the Soviets always plan and build redundant systems wherever possible. Third, the Baranovichi radar lies on a straight line between Moscow and U.S. Pershing II deployment areas in West Germany, which is consistent with the Soviets' expressed concerns about the Pershing as a quick-strike, anti-C^3 weapon. Baranovichi could be tasked to concentrate on the Pershing "threat tube," a relatively narrow azimuth band, permitting a more rapid search rate and closer tracking of incoming warheads. The neighboring radars could concentrate on early warning of ICBM and SLBM attack. (Baranovichi is probably a poor site for a battle management radar for anti-Pershing defenses, however, since it is located too far forward to track Pershing warheads targeted on the Soviet interior after they reenter the atmosphere, a critical shortcoming since, unlike other U.S. ballistic missiles, the Pershing II has active terminal guidance and maneuvers upon reentry.)

Rapidly-Deployable ABM Components. Since the mid-1970s, the United States has kept track of a half-dozen or so small (by Soviet standards) ABM radars, designated "Flat Twin" and "Pawn Shop" by NATO and located at Soviet ABM test sites. Flat Twin reportedly can be assembled on prepared sites in a few months; in 1975 it was observed to have been disassembled and relocated (from Sary Shagan to the Kamchatka peninsula) in a matter of months.[52] The van-mounted Pawn Shop radar is a modular unit but apparently has never been observed to have been moved. In the late 1970s, the United States estimated that ABM sites making use of these radars could probably be set up in about six months, in contrast to the sites at Moscow, which required years to construct and additional years to upgrade. At the time, neither radar was considered a "mobile" ABM component, "in the sense of being able to be moved about readily or hidden."[53] More recently, a 1986 U.S. report on Soviet arms control compliance stated that mobile ABM components include those "which can be readily transported from one place to another as well as components designed to be moved frequently during their service life." That report concluded that evidence on the Flat Twin and Pawn Shop radars is "ambiguous"; nonetheless, they are a "potential violation" of the Treaty.[54] A 1974 explication of Treaty limits written by John Rhinelander, the former legal advisor to the U.S. SALT delegation, argues that the prohibition on mobile land-based components applies both to fully mobile components and to those that are merely transportable, because the latter do not qualify as "permanent, fixed types" as stipulated by Common Understanding C.[55] By that reading, both these radars are impermissible.

The Soviet Union may have taken steps in the fall of 1986 to reduce or eliminate this issue. Of the 6 radars built (apparently 2 Flat Twin and 4 Pawn Shop), 5 were located at the Sary Shagan site in Soviet central Asia. "Most or all" were dismantled. The Kamchatka Flat Twin remains.[56]

George Schneiter, former Deputy Director of the Defense Department SALT Task Force, notes that the Treaty actually is silent on the question of "rapid deployability," which is as much a source of U.S. concern over the Flat Twin and Pawn Shop radars as their potential transportability. A party is legally free to design components to be rapidly deployable, so long as those components, once deployed, are permanently fixed in place.[57]

Even if the Soviet Union could deploy a Flat Twin radar in a matter of months, deploying a significant number of units would take much longer. The postulated rapid-deployment system would, moreover, use above ground launchers for Sprint-like interceptors (unlike the upgraded Moscow system, which uses silo launchers), and those would be quite vulnerable to nuclear blast.[58] The Flat Twin and Pawn Shop radars are, moreover, terminal tracking and guidance radars, not battle management radars.[59] A number of radars similar to the new LPAR at Moscow (recently dubbed "Pill Box" by NATO) would be needed for battle management, and they take several years to build. No evidence of their construction has yet come to light.

Rapid-Reload ABM Launchers. The Reagan administration's compliance documents also highlight the reload capabilities of Soviet ABM launchers, suggesting that they approach a prohibited "rapid reload" capacity. A Galosh ABM test launcher at Sary Shagan reportedly was reloaded in "hours" in 1983, and an SH-8 launcher at the same test site reportedly launched two missiles in two hours. The apparent lack of reloading equipment at the site between first and second firing has led some to infer an underground reload system for the SH-8 silo.[60] (Such a reload system would clearly violate the treaty, but on-site inspection of randomly selected silos could settle the issue.)

While the Treaty was being negotiated, U.S. Chief Negotiator Gerard Smith told his Soviet counterparts that the United States considered "rapid" to mean "in a strategically significant" period of time. At that time, the United States credited the Galosh ABM launcher with a 15 to 30 minute reload capability.[61] There is no indication in reports on Soviet compliance issued in the late 1970s that the United States challenged that reload capability before the SCC, although a number of other Soviet activities were challenged. U.S. criteria defining rapid reload capability may have changed since then. A two-hour reload capability is, however, not all that rapid; reload crews and missiles would be exposed to follow-on attacks in those two hours.

Exempt Systems: Air Defenses and ATBMs. Since the Treaty only constrains "systems to counter strategic ballistic missiles ...," air defense and anti-tactical ballistic missile systems (ATBMs) may be developed and deployed in any configuration and in any number, provided they do not have capability to counter strategic ballistic missiles and are not "tested in an ABM mode."[62] The ATBM exception protected U.S. options to deploy an ATBM; at the same time, these provisions attempted to confine non-ABM systems to less than full ABM capability. Known then as the SAM upgrade problem, this was of particular concern to the United States in SALT. After the Treaty was in force, indications that Soviet air defense radars associated with the SA-5 SAM were being switched on in conjunction with strategic ballistic missile flight tests led to challenges before the SCC and to two still-secret protocols (of 1978 and 1985) governing testing in an ABM mode.[63] U.S. concerns about such threats "from below" to the integrity of the Treaty are thus of long standing. Current issues involve Soviet testing and deployment of new mobile SAM/ATBM systems, and the tactical ballistic missile threat to NATO, to which ATBM is one possible response.

There are actually three related policy questions to be answered: Can ATBM technologies and deployments undermine the Treaty? Does NATO need ATBM? And if so, can verifiable distinctions be made between ABM and ATBM capabilities? If technical requirements for ATBM and ABM are *dissimilar* and NATO needs ATBM, then the United States might seek agreed-upon definitions of ATBM that set a cap on its capability—to prevent erosion of the Treaty—without discouraging its development and deployment. On the other hand, if the technical requirements of ATBM and ABM prove to be indistinguishable, there may be a problem unless ATBM turns out to be unimportant for NATO, because preventing erosion of the Treaty would then argue for a ban on ATBM.

Threats to the Treaty. As air defense technology continues to improve, ATBM and SAM upgrade issues may pose serious threats to the integrity of Treaty limits—if not in technical fact, then in political perception. (The Soviet SA-X-12B GIANT is an example of such a perceived threat.[64]) It is not so much that ATBMs could provide a damage-limiting defense of national territory—that is difficult enough to achieve with systems designed explicitly for BMD. Rather, the side observing their deployment may not be able to determine with confidence that ATBMs could *not* take a significant toll of strategic missile RVs (deployed as terminal defenses of military targets, for example), or that their deployment did not reflect a conscious effort by their deployer to "creep out" of the Treaty. On the other hand, mobile ATBMs would be inherently soft targets for nuclear attack; likely intercept probabilities that are much lower than regular BMD could add up to an intercept capability more apparent than real. Nonetheless, if perceptions of threat from ATBMs are not to undermine the Treaty, then some sort of constraints may be in order.

NATO Requirements. Whether NATO needs ATBM capability is a subject beyond the scope of this discussion, but a few words are in order. The Soviet Union is deploying several classes of tactical ballistic missiles (TBMs) in Europe, reportedly able to be equipped with non-nuclear warheads (e.g., chemical or improved conventional munitions) suited to attacking NATO air bases, command and control facilities, nuclear storage sites, supply depots, and similar targets of high value early in a NATO-Warsaw Pact conflict. The less accurate predecessors to these TBMs were assumed to be primarily nuclear-tipped. NATO's own nuclear capability would have been the principal deterrent to their use and, in any case, the binary nature of the soft-target nuclear defense problem (perfect defense or no defense) meant that effective defense of unhardened NATO targets against ballistic missile attack was not feasible. Against non-nuclear warheads, however, an imperfect defense might still contribute to NATO's ability to maintain C^3 and to muster aircraft against an initial Warsaw Pact attack.

In a recent study, Benoit Morel and Theodore Postol examine the TBM threat to NATO, the technical requirements for anti-tactical missile defense, alternatives to it, and alternative air threats to NATO. They conclude that TBMs are a minor component of the overall Warsaw Pact air threat to NATO; that other vehicles, especially cruise missiles, could perform the same preemptive functions as TBMs and do so much more cost effectively; and that measures other than ATBM could contribute more effectively to NATO force and C^3 survivability against TBMs and the other threats. Such measures include hardening, dispersal, concealment, mobility, construction of additional airbase runways for emergency use, and increased reliance on vertical/short take off and landing (VSTOL) aircraft.[65] Based on progress in negotiations as of spring 1987, arms control may also greatly reduce the TBM threat, although current arms control proposals would not reduce the other threats that Morel and Postol believe constitute the real time-urgent non-nuclear air threat to NATO.

However, if NATO did decide to develop, test, and deploy ATBM, the Treaty would have something to say about how the United States went about contributing to that program and what U.S. technologies could be involved. Article IX forbids the deployment of ABM systems or components by either party outside its national territory, and forbids the transfer of ABM systems and components to third parties. Agreed Statement (G) associated with Article IX also forbids the transfer to third parties of "technical descriptions or blueprints specially worked out for the construction of ABM systems and their components limited by the Treaty." The Statement is worded so as to stem proliferation of ABM technology.

Some technologies currently slated for development under SDI as strategic interceptors (for example, the Low Endoatmospheric Defense Interceptor [LEDI] or the Flexible, Agile Guided [FLAG] Experiment) might, as terminal defenses, be suited to an ATBM role. If either were tested in an ABM mode, however, it could only be deployed as permitted by Article III, and the technology would be subject to the no-transfer provision of Article IX. The same would be true for airborne long-wave infrared sensors, which may be needed to give crucial "cues" to ground based ATBM tracking radars.

Distinguishing ATBM from BMD. Herbert Lin suggests that clear distinctions might be drawn based on the performance capabilities of SAM/ATBM interceptors and radars, and thus that verifiable limits might be drawn between systems designed for such purposes and systems to counter strategic ballistic missiles.[66] Morel and Postol suggest that the difficulty of intercepting TBMs may be greater than generally assumed; that systems such as Patriot may have great difficulty dealing with the TBM threat without the assistance of large search and acquisition radars or long-range, airborne infrared sensors, both of which resemble ABM components.[67] Augmented ATBM systems thus may not be distinguishable from ABM systems; on the other hand would-be ATBMs without access to such external cueing would be much less capable than an ABM system.

Breaking Out Is Hard to Do. All the foregoing issues, combined with the absence of direct constraints in the Treaty on ABM component production, create worries about "breakout"—rapid, clandestine, or otherwise unannounced steps to deploy BMD quickly in an effort to gain a strategic advantage. Since this is a recurrent theme in the BMD debate, it is worth examining more closely. What might constitute meaningful breakout and what might the perpetrator gain by it? A meaningful breakout would be one that fielded a complete defense system before the other side could neutralize it by means of offensive countermeasures, defensive counter-deployments, or some combination. Although the U.S. Defense Department frequently speaks about the threat of Soviet breakout, not even Defense Secretary Caspar Weinberger doubts the ability of the United States to pace Soviet BMD deployments with U.S. deployment of offensive countermeasures.[68] These would include penetration aids (such as decoys) and accelerated development and deployment of maneuvering reentry vehicles and cruise missiles. If offensive countermeasures kept pace with a breakout effort (or if Soviet decisionmakers *believed* that they kept pace), the system would never be exploitably complete.

What might the owners do with such a breakout advantage if they managed to acquire one? From the standpoint of Soviet doctrine, they might do little more than work to keep a strategic edge, once obtained, since an advantageous "correlation of forces" is considered to be a guarantor of peace (and influence). If efforts to influence took a coercive turn and if the breakout system were built on currently available defense technologies, the West could—to borrow a phrase from the Reagan social program—"just say no." BMD would not render the Soviet Union immune to Western weapons. But this is clearly not a situation that the West should allow to occur.

Indeed, Western deterrence doctrine, coupled with standard worst-case assumptions about Soviet political/military intentions, would paint a much darker picture of Soviet breakout: Any advantage very likely would be transient and thus, to be useful, it would have to be quickly exploited following system completion. Yet only if East-West relations had already deteriorated so badly that there were, as Khrushchev once put it, a "smell of burning in the air," and if Moscow were perceived to be on the verge of achieving a very significant BMD capability, *and* if its air defenses simultaneously approached damage-limiting capability against U.S. bombers and cruise missiles (that is, increased expected bomber/cruise attrition from 25 percent to between 80 and 90 percent) could Western options be perceived to have narrowed to a very small and unpleasant set.

Keeping East and West out of such cul-de-sacs is the task of diplomacy, arms control, and military hedging behavior. The United States has taken precisely the right approach in developing a variety of offensive countermeasures that could be deployed quickly. In comparison to a counter-deployment of BMD, these measures are relatively cheap and more conducive to maintaining a stable Treaty. A strategy that hedges against Treaty breakout by developing a near-term BMD deployment capability may encourage what it seeks to discourage, since a defense-oriented hedge may be indistinguishable to the other side from a breakout strategy, and can give the other side a rationale for walking away from the Treaty. The Soviet Union seems to have such a defense-oriented hedging strategy. According to some U.S. officials it is indistinguishable from a breakout strategy and does give them a rationale to advocate abandoning the Treaty.

Development Issues

This category treats issues that do not involve current hardware deployments. They may seem to have less immediate or tangible impact on the integrity of the Treaty, but they do impinge directly on SDI. Definitions of such terms as "component," "development," and "adjunct" set the legal and political limits of SDI, as well as the limits of comparable Soviet programs.

If the Treaty were reinterpreted so as to remove these limits on the development and testing of new ("exotic" or "non-traditional") BMD technologies, SDI could have somewhat smoother sailing, although independent assessments suggest that physics and engineering constraints pose as formidable a barrier to SDI development as do politics and diplomacy.[69] Important elements in the Reagan administration favor the reinterpretive approach as a sophisticated alternative to withdrawal from what some openly describe as a bad bargain for the United States. Because the interpretation issue affects other Treaty-related development questions, it is addressed first.

The New Interpretation. In October 1985, the Reagan administration announced a new ("broad" or "expansive") interpretation of the Treaty, under which the development and testing of exotic ABM systems and components would be permitted without restraint—principally to clear the way for testing SDI components in earth orbit (all parties agree that the Treaty permits testing and development of exotics that are fixed land-based). The following material reviews the origins and reasoning behind this new interpretation, reactions and assessments of it, and its role in the Reagan administration's arms control policy.

Origins and Reasoning. The ABM Treaty presents problems for those who see it as a temporary and artificial barrier to development and deployment of a very effective BMD. Systems and components with basing modes thought to offer the greatest promise of improved defense effectiveness are those whose development, testing, and deployment are banned under Article V (namely, those involving mobility of kill platforms and sensors). If technologies cannot be tested in their proper basing mode, their potential effectiveness cannot be adequately assessed; if their effectiveness cannot be assessed, there is great risk in breaching the Treaty in order to deploy them—subsequent testing might uncover a fatal flaw in the system, leaving the United States without a workable defense and Soviet BMD without restraint.

By early 1985, as indicated in its report to Congress, SDI Organization (SDIO) was coping with this problem by structuring its test and demonstration program in ways that would "work around" Treaty limits. Some technologies were to be tested at less than their full operational potential, while others were to be tested with key subcomponents omitted. Certain space-based equipment would be tested only against co-orbital targets whose speed and trajectory relative to that of the test equipment would not have the characteristics of a strategic ballistic missile or reentry vehicle. Many of the proposed tests were more costly and time-consuming than would be a straightforward test of the technology.[70]

In May 1985, a query during Senate hearings on Donald Hicks' nomination as undersecretary of defense for research and engineering (USDRE) led a former CIA lawyer on Iklé's staff to assess the impact of the ABM Treaty on SDI. That effort led to a more extensive review by the same individual of the SALT I negotiating record the following September. No members of the U.S. SALT I delegation were interviewed.[71]

The report that subsequently emerged from the office of Assistant Secretary of Defense Richard Perle in late September 1985 averred that non-traditional ABM systems and components (those based on "new physical principals") were not constrained *at all* by the Treaty, because the Soviets during the negotiations had refused to accept such limits. The Office of the Legal Advisor in the State Department undertook its own review of the negotiating record and basically confirmed Perle's findings. (Special Advisor to the President on Arms Control Paul Nitze was the only member of the U.S. SALT I delegation consulted in the course of the review. None of the other delegation members who subsequently spoke on this issue agrees with this reinterpretation of the Treaty.) However, the State Department review did read the Treaty as banning the deployment of non-traditional BMD. One day after the legal advisor, Abraham Sofaer, reported his findings to Secretary of State George Shultz, the Special Arms Control Policy Group convened at the White House to consider those findings. The group adopted the new interpretation and reportedly decided to offer Moscow five- to seven-years' notice of intent to withdraw from the Treaty. This decision was reflected in a Reagan letter to Gorbachev the following summer. The new interpretation surfaced sooner, during an appearance on *Meet the Press* by National Security Advisor Robert McFarlane.[72]

The resulting political furor, especially among the NATO allies, within a week led to another high-level White House meeting on the Treaty. There, Secretary of State Shultz reportedly succeeded in convincing Kenneth Adelman, Director of the Arms Control and Disarmament Agency, Defense Secretary Weinberger, McFarlane, and Reagan that the United States should continue to uphold the traditional ("strict" or "restrictive") interpretation of the Treaty as a matter of policy, but not as a matter of legal obligation.[73]

Sofaer testified before Congress on the new interpretation. That testimony was revised and published in the *Harvard Law Review*.[74] Sofaer's arguments against the traditional interpretation follow three paths: The Treaty language, on its face, is ambiguous; the United States tried but failed to obtain Soviet support for banning exotics; and the post-negotiation public record is ambiguous as to U.S. government policy on the matter.

Sofaer's first point is that Article II(1) of the Treaty is not a functional definition of an ABM system and its components that merely uses traditional components as an example, but rather is a precise definition of what elements the Treaty is intended to cover. Since only launchers, radars, and interceptors are named, only those components are constrained.[75] The only instance in which the Treaty mentions components based on "new physical principles" is in Agreed Statement D associated with Article III (which defines the exceptions to the Treaty's overall deployment ban). That Statement, Sofaer notes, "explicitly allows the 'creation' of such systems and components; it requires that limitations on such systems be stipulated only *after* creation of the systems. . . . Nothing in Agreed Statement D, however, states that it applies only to future systems that are fixed land-based."

Sofaer also concludes that Agreed Statement D was "the furthest the Soviets were willing to go" on exotics. He argues that "U.S. negotiators persuaded the Soviets to adopt Agreed Statement D by explaining that without it, the Treaty would leave the parties free to *deploy* future systems or components based on other physical principles."[76]

Finally, Sofaer contends that the U.S. government itself did not unequivocally adhere to the traditional interpretation of the Treaty. He cites a number of statements by U.S. officials regarding the Treaty's impact on development and testing of exotic technologies that do not explicitly distinguish between fixed, land-based technologies and others. He concludes that this at best reflects ambivalence in the U.S. position.[77]

Incremental Implementation. Shortly after the October 1985 endorsement of the broad interpretation by the White House, Perle characterized the administration's decision to abide by the traditional interpretation as "temporary."[78] In Senate hearings the following spring, Perle described the new version as "the only legal" interpretation of the Treaty and predicted that the administration would see that there was "no rational basis for long-term adherence to the restrictive interpretation."[79]

Soviet reaction to the initial 1985 announcement was both swift and negative. Writing in *Pravda*, Chief of the Soviet General Staff Marshal Sergei Akhromeyev affirmed the traditional reading of the Treaty, stating that "research, development and testing of ABM systems or their components based on other physical principles is allowed in areas that are strictly limited by the treaty and defined by it, and only using permanent ground [-based] ABM systems." In May 1986, Soviet negotiators in Geneva proposed that the two sides agree on measures to strengthen the Treaty, and pledge not to exercise their right to withdraw for 15 to 20 years; informally, Soviet officials indicated that a 10- to 15-year pledge would be acceptable.[80]

In July 1986, Reagan replied in a letter to Gorbachev by offering a five-to-seven-year period of Treaty observance governed by the broad interpretation, followed by freedom to deploy BMD. Reagan placed the same proposal on the table at the October 1986 Reykjavik summit, lengthening the period of observance to ten years. The Soviets in turn proposed that, for a period of ten years, both the United States and Soviet Union "adhere strictly" to the provisions of the Treaty.[81]

Although the two sides' proposals were superficially similar, their goals were very different. Where Moscow was looking to lock into the Treaty for *at least* another decade, with tight constraints on space testing, Washington was looking to abide by the Treaty for *at most* another decade, with lessened constraints on space testing of exotic technologies, and with deployment the goal at the end of the period. In short, the United States sought to change the Treaty from one of indefinite duration to one with a non-renewable ten-year limit. At a news briefing shortly after the summit, National Security Advisor John Poindexter confirmed that the administration was offering to delay U.S. withdrawal from the Treaty in exchange for Soviet agreement to the broad interpretation.[82]

Having failed to influence the Soviet position at Reykjavik, the Reagan administration reassessed its options. At home, on the technical front, a comprehensive report on directed energy weapons (lasers and particle beams) by the American Physical Society was undergoing Defense Department classification review. The report, finally released in April 1987, dismissed the near-term weapons potential of directed energy technologies, concluding that at least another decade of research was needed to reach an informed decision about whether these technologies would *ever* be worth pursuing as weapons.[83] Directed energy weapons were initially the main focus of SDI research and were seen as key to fulfilling Reagan's dream of effective defense.

The programmatic emphasis of SDI already had been shifting from space-based lasers and particle beam weapons, to ground-based lasers and orbital mirrors, to magnetically-driven, hypervelocity "railguns," and the shift continued in late 1986. The shift reflected in part an effort by program managers to stick to the program's original milestone of a deployment decision by 1993–94 in the face of lower-than-requested budgets and receding feasibility estimates for the more exotic systems. In part, it also reflected the concern of politically conservative supporters of SDI that funding could not be sustained or increased if the program focused largely on technologies having only very long-term payoff. These concerns were grounded in political reality. Although the fiscal 1987 Congressional authorization for SDI was substantially higher than that of the previous year, it was about one-third less than the administration had requested. Thus, in fall 1986, conservatives began to argue in favor of a near-term deployment option for SDI.[84]

In December 1986, Reagan was briefed by Weinberger, SDIO chief General James Abrahamson, and Perle on a revised approach to SDI, an option for early 1990s deployment of a "phase one" system using only homing rocket interceptors based on land and in space. The phase one system strongly resembled that recommended by the private group High Frontier, whose plans formerly had been strongly criticized by the Reagan administration, including officials of SDIO.[85]

In mid-January 1987, before the Senate Armed Services Committee, Weinberger advocated moving to phase one deployment of an eventual nationwide BMD system "as quickly as possible." On February 3, Reagan convened the Cabinet-level National Security Planning Group to discuss an early deployment decision and adoption of the broad interpretation to facilitate SDI testing, particularly with regard to an upcoming test of orbital rocket interceptors (the so-called Delta 181 test). Reagan is reported to have favored its adoption without public announcement. Word of the meeting prompted a warning from Senator Sam Nunn, the moderate-to-conservative chairman of the Senate Armed Services Committee, who was then preparing his own reports on the broad interpretation, that any such move would lead to a "Constitutional crisis." Decisions on both early deployment and the broad interpretation were postponed. A deployment decision, according to Shultz, was about two years away.[86]

Interpretation of the Treaty also has been an issue at the Geneva arms talks. In mid-January 1987, U.S. Chief Negotiator Max Kampelman agreed with his new Soviet counterpart, Yuli Vorontsov, to alter the format of the talks, shifting from stiff, formal sessions to a less formal working group format. One working group was to discuss limits on development and testing permitted under the Treaty. Recalled to Washington two weeks later, Kampelman was upbraided by Perle for exceeding his instructions in Geneva. Shortly thereafter, the U.S. delegation reportedly "repeated" an offer that the United States would abide by the narrow interpretation during "any mutually agreed period of non-withdrawal from it." This delegation action seems to have prompted new instructions from the White House forbidding discussion of development limits that were inconsistent with the broad interpretation. The Soviets publicly complained of a U.S. "shift" to the broad interpretation, which the U.S. delegation promptly and publicly denied, taking pains to assert that the narrow interpretation remained U.S. policy. Such unusual public action was further indication of severe stress between Washington and the delegation.[87] Two weeks after this exchange, Senator Nunn released his reports on the new interpretation.

Reactions and Analysis. The broad interpretation has been sharply criticized both on its merits and for its impact on the Treaty. Reactions were particularly sharp after the Reagan administration moved toward an early deployment decision—with notable reactions from NATO allies, former U.S. arms negotiators, six of the last seven Secretaries of Defense, and from Capitol Hill.[88]

On Capitol Hill, U.S. Senators sought and received access to the SALT I negotiating record to assess for themselves whether that record could reasonably be read as supporting the broad interpretation, and to relate the classified negotiating record to what the Senate had been told in the 1972 ratification hearings. Senator Carl Levin was the first to complete a review of the record in late 1986; he condemned the new interpretation unequivocally.[89] Three months later, Senator Nunn reached similar conclusions. Speaking on the Senate floor over three consecutive days, he charged that the administration's account of the ratification record was incomplete and that based on that record its case for reinterpretation was non-existent; that it had presented no evidence confirming either U.S. or Soviet support for a broad interpretation of the Treaty in practice; and that the evidence gleaned from the negotiating record by Sofaer was "woefully inadequate" to support the case for reinterpretation. Nunn warned, moreover, that the issue involved more than a difference of opinion over Treaty language. The State Department, he said:

> has argued that regardless of whether the ratification proceedings support the Reinterpretation, Executive Branch testimony presented to the Senate during the treaty-making process can be disregarded because it has "absolutely no standing" with the Soviets. . . . The State Department is arguing, in effect, that Administration witnesses need not accurately reflect the Executive's understanding of a treaty; instead, they are free to keep that understanding a secret and may mislead the Senate into consenting to a treaty which has a secret interpretation different from the meaning presented to the Senate. . . . [B]y asserting that Executive Branch statements to the Senate are essentially meaningless, the State Department is risking a serious constitutional confrontation. . . . The effect may well produce a Congressional backlash through exercise of the power of the purse and the power to raise and support armies in a manner that would give effect to the original meaning of the Treaty as presented to the Senate.[90]

The thesis that Nunn rejected could apply, of course, to more than the ABM Treaty. The first steps in the predicted backlash were evident in April 1987, when the House Armed Services Committee included language in the 1988 Defense Authorization Bill that required adherence to the strict interpretation of the Treaty.[91]

Let us look at the substance of the interpretation issue in greater detail, in terms of U.S. and Soviet policy and practice. Article II's definition of an ABM system as "currently consisting of" ABM interceptors, launchers, and radars generally has been taken to mean that the Treaty applies, and was intended to apply, to more than the cited means of countering strategic ballistic missiles; that the signers anticipated the evolution of technology; and likewise that they sought its limitation. U.S. policy during the negotiations (as set by National Security Decision Memorandum 127 of August 12, 1971) sought a ban on deployment of all exotic ABMs and a ban on development of all but fixed, land-based exotics. Careful reviews of ratification hearings moreover confirm that this was indeed both the Executive and the Senate's understanding of the Treaty at the time of ratification.[92] No revised understanding was developed between then and 1985. Although within the U.S. defense community there has always been a minority view that the Treaty's writ did not extend to exotics, not until the second Reagan administration did that view ever come close to being prevailing orthodoxy. Indeed, in his review of the historical record, Nunn was unable to locate any statements by U.S. officials prior to October 1985 that explicitly avowed the new interpretation.[93] Administration Arms Control Impact Statements (ACIS), which have been required by Congress since the mid-1970s and reflect interagency consensus on arms control policy issues, explicitly endorse the traditional interpretation through the edition issued in April 1985. Through that edition (for fiscal 1986), the ACIS concur that the Treaty covers "exotic" ABM technologies; that their development and testing in a fixed, land-based mode is permitted; and that their deployment is prohibited unless the two parties meet and agree upon how they can be deployed in a fixed, land-based mode consistent with Article III. The fiscal 1987 ACIS, issued in April 1986, is the first departure from this position. It endorses the revised interpretation as "fully justified," but states that the older interpretation will be observed, budgets permitting.[94]

The Soviet Union was largely silent on the interpretation question until 1985. Supporters of the Sofaer position argue that this reflected Moscow's reservations about limits on exotic BMD. On the other hand, Raymond Garthoff, who participated in the SALT I negotiations on future ABM technologies, is emphatic that Soviet agreement was obtained on the Treaty's application to future technologies, and that the word "currently" was inserted ahead of the list of components in Article II in December 1971 to ensure that the Treaty covered more than that list. Agreed Statement D then was drafted to ensure that the Treaty was seen to permit the development and testing of fixed, land-based exotics even as it permitted the development and testing of other fixed, land-based ABM systems and components. The Soviet Union accepted the Agreed Statement in late January 1972. Only later that spring were the deployment provisions of Article III put in final form, once both sides had agreed upon the type and number of ABM deployment areas to be permitted.[95]

Critics of Sofaer's approach also note that silence is a form of consent; that Soviet failure from 1972 to 1985 to gainsay U.S. government statements on the meaning of the Treaty would have given the United States good grounds to object under international law if it were Moscow that sought to test space-based ABM technologies. Consistent with this argument, when the Soviets were confronted with a Treaty interpretation by the United States to which they objected, they did so vigorously. Since 1985, statements by high Soviet Party and military officials have put Moscow on record as favoring the traditional interpretation. Thus, if the problem with interpreting the Treaty revolved around doubts about Soviet attitudes to it, the United States could readily lock in Soviet support for the traditional interpretation and erase any doubts derived from an incomplete negotiating record, through the Geneva Nuclear and Space Talks or through the SCC.[96]

On the question of U.S. and Soviet practices since 1972, Nitze, who favors the broad interpretation, testified that as of October 1985 neither the Soviet Union nor the United States had violated the restrictive interpretation of the Treaty.[97] A closer look at U.S. and Soviet BMD programs since 1972 substantiates that observation. Following Congress's restructuring of the BMD program in the mid-1970s, the United States maintained a steady-state program of research geared to hedge against technological surprises (including Soviet "breakout") and to improve U.S. offensive systems' ability to penetrate potential Soviet defenses. Other efforts outside the U.S. Army's BMD program pursued directed energy technologies, but largely for purposes other than BMD. One may argue whether such efforts were adequately funded, but it is difficult to argue that U.S. programs during the first decade of the Treaty's existence suggest exploitation of anything like the broad interpretation with respect to development of exotic BMD.[98]

The Soviet BMD program, although harder to document, probably has been more broadly based than similar U.S. research in the past fifteen years. By the early 1980s, the Soviets had begun to modernize the Moscow ABM system into a near clone of Safeguard. The deployment has been incremental, in the tradition of the Soviet weapons establishment. Although there are large ground-based lasers at the ABM test site at Sary Shagan, and considerable particle beam work, current U.S. concerns that the Soviet Union is moving to "provide a base" for BMD of national territory focuses not on these exotic technologies but on variants of traditional technologies.

In sum, neither U.S. policy before October 1985, nor Soviet policy, nor U.S. and Soviet practices can be shown to support any other conclusion than that the Reagan administration's broad interpretation was an abrupt departure in policy. That it conflicts with the Treaty's basic purpose—to forestall the deployment of nationwide BMD—is clear. Under the broad interpretation either side would be free to test, for example, land-mobile lasers or space-based railguns in any quantity and in any locale. The broad interpretation was only intended to facilitate SDI, and to do so in a manner that could also put later administrations on an irreversible course toward SDI deployment. There are two ironies in this: One is that the new interpretation probably was not needed, since the Defense Department's previous "work around" strategy seemed able to accommodate most of SDIO's experimental requirements for a number of years; it only made more visible a hardline strategy that drew criticism from such key potential allies as Senator Nunn. Second, SDI's fallback strategy, based on rocket interceptors, may not be implementable under *any* interpretation of the Treaty. Rockets are a traditional form of BMD, even if their launchers orbit the earth instead of hugging the ground.[99]

However, the status of space-based rockets is only one of a number of development-related challenges currently facing the Treaty. Others are discussed below.

Defining "Component" and "Develop." In the SALT I negotiation, the United States sought to limit possible Soviet circumvention of the Treaty via deployment of large numbers of ABM-capable radars without launchers and interceptors (that is, not as part of an "ABM system"). Thus, the United States proposed limits on ABM "components" as well as complete systems. The term is reasonably clear when applied to those ABM components (radars, launchers, and interceptors) that are explicitly defined in the Treaty, although labelling even a familiar type of technology as an ABM component is not always straightforward. With respect to some of the newer technologies having ABM potential, such labelling could be more difficult, both because some of these technologies distribute ABM functions in novel ways, and because there are no agreed-upon standards about what gives a technology "capabilities to counter" a strategic ballistic missile (language found in Article VI) and what makes it "capable of substituting" for a traditional ABM component (language found in Agreed Statement D).[100]

The limits on "development" contained in Article V and elsewhere in the Treaty were added at U.S. insistence, to reinforce limits on component testing. Thus, if U.S. (or Soviet) NTM were to spot suspicious activity, the Treaty sanctions questioning such activity even if no tests have been actually observed (because testing might not have begun or might have been missed by NTM).[101]

Since the Soviet Union and the United States have followed somewhat different paths on these concepts, it is worth reviewing their respective positions before discussing related problems for the Treaty.

Soviet Positions. The Soviet-language version of Article V uses somewhat more restrictive wording than does the English version. In the former, both sides agree "not to create" (*ne sozdavat*) the types of systems or components prohibited by Article V, a broader usage than what the United States meant by "develop." This helps to explain Soviet insistence, beginning in 1985, that the Treaty bans even basic research into such areas as space-based ABM components.[102] That position was reflected at the Reykjavik summit, where Gorbachev proposed that "Testing of all space elements of anti-ballistic defense in space [*sic*] are prohibited except research and testing in laboratories."[103] After Reykjavik, the Soviets offered a series of clarifications. Director of the Soviet Institute of Space Research Roald Sagdeev suggested that some tests in space might be allowed, as "we scientists consider manned space stations as orbital laboratories." (The Soviets, it should be noted, currently have a permanently inhabited station, *Mir,* in earth orbit.) Moreover, according to Sagdeev, joint assessments might determine whether a given test contravened Treaty limits and also establish capabilities thresholds beneath which technologies could be freely tested.[104] At a November 1986 news conference following talks with Secretary Shultz, Soviet Foreign Minister Eduard Shevardnadze offered further clarification: "Laboratory research" could include "the building of ready samples, prototypes of corresponding defensive systems...."[105] As noted earlier, the White House curbed discussion of Treaty limits at Geneva in early 1987; when Shultz went to Moscow in mid-April, the Soviet Union again requested talks on the subject, suggesting that the two sides draw up a list of items that could legally be launched into space.[106]

The Soviet approach to defining Treaty limits is consistent with the division of Soviet military research and development into scientific-research work (*nauchno-issledovatel'skaya rabota* [*NIR*]) and experimental-design work (*opytno-konstruktivnaya rabota* [*OKR*]). *NIR* "involves concept studies and proof of principle research to support the creation of new technologies. . . . [T]he Soviet *NIR* process is heavily test oriented; the Soviets build demonstrators and mockups where western research institutes use computer simulations." But it does not necessarily feed directly into weapons engineering and production. The ground-based laser at the Sary Shagan ABM test facility may be such a feasibility demonstrator. In twenty years of *NIR* on laser weapons, none appears to have shifted to *OKR*, which is the process of actual "development of a weapon system intended for serial production."[107] *OKR* roughly corresponds to what the United States would call engineering development.

U.S. Positions. The U.S. understanding of the term "develop" as used in the Treaty, particularly in Article V, was publicly defined by Ambassador Smith before the Senate Armed Services Committee on July 18, 1972. It remained the official U.S. definition through late 1986, as follows:

> The prohibitions on development contained in the ABM Treaty would start at that part of the development process where field testing is initiated on either a prototype or breadboard model. It was understood by both sides that the prohibition on "development" applies to activities involved after a component moves from the laboratory development and testing stage to the field testing stage, wherever performed. The fact that early stages of the development process, such as laboratory testing, would pose problems for verification by national technical means is an important consideration in reaching this definition. Exchanges with the Soviet Delegation made clear that this definition is also the Soviet interpretation of the term "development."[108]

By the Smith definition, then, Treaty limits stop where NTM monitoring stops, and apply to NTM-observable field tests of prototype or pre-prototype ("breadboard") components that have ABM capability (that is, the ability to assume the functions of an ABM radar, ABM launcher, or ABM interceptor in countering a strategic ballistic missile or its elements "in flight trajectory").

In October 1986, two weeks after the Reykjavik summit, the Reagan administration clarified its position on "research" and "development" in an address by Nitze. Mr. Nitze defined research as including "conceptual design and testing conducted both inside and outside the laboratory." Development "commences with the construction or testing of one or more prototypes of the system or its major components." Nitze's definition also hews to the broad interpretation, claiming allowance for space testing of ABM systems and components "based on other physical principles."[109]

Problems. On the one hand, if "component" really refers only to something that is, or is capable of performing the function of an ABM radar, launcher, or interceptor, then Article V can be construed as applying only to those test items that are already more or less fully developed. If the development of an item cannot be challenged until such development is virtually complete, then the function of Article V as a buffer against ABM Treaty breakout is clearly undermined.

On the other hand, applying the Treaty to technologies that have some ABM potential but have not yet demonstrated an ABM capability (e.g., by tracking strategic reentry vehicles, intercepting an ICBM booster, or serving as a launch platform for such intercepts) would extend testing limits into technologies that one side may not want to constrain.

Further, there is the question of an ABM component "adjunct"— something that could help an ABM system do its job, without being critical for the system to work, and without being capable of functioning as a component itself. Thus, Rhinelander notes that "Certain devices, such as telescopes, which are simply adjuncts to, not substitutes for, present ABM components are not covered" by the Treaty. And Garthoff notes that "the 'capable of substituting for' language" in the Treaty "was crafted to allow, for example, laser adjuncts to other components for missile detection or ABM missile guidance but to bar lasers as interceptors or as substitutes for radars."[110] Where adjunct status ends and component status begins is a matter of debate.

An example of an adjunct at the high end of the capabilities curve is the Airborne Optical Adjunct (AOA), an Army project funded under SDI to develop a large-aperture airborne infrared telescope.

The telescope is a passive sensor (one that tracks the target's naturally-emitted radiation, in this case, heat; active sensors such as radars track their targets via echoes of radiation deliberately bounced off the target). The planned operational version, the Airborne Optical System (AOS), is intended to pass on tracking data to a ground-based ABM radar and thus to remain an adjunct, albeit a much more capable one than Treaty negotiators may have had in mind.[111]

If AOS were able to pass on data directly to incerceptors, thereby performing the functions of an ABM radar, in all likelihood it would be considered an ABM component, and as an "air-based" sensor its development, testing, and deployment would be banned. This raises the following question: Must a radar or sensor be able to perform *all* functions, from target acquisition to terminal tracking and guidance, to be considered an ABM radar for purposes of the Treaty? The historically-minded point out that Nike Zeus used as many as 4 radars in sequence; that the Moscow system uses 2 and possibly 3, as would the so-called ABM-X-3 architecture; and that Safeguard used 2 radars. Subsequent U.S. terminal defense system designs have relied on just one radar "layer," the only external cues coming from distant early-warning systems. If technologies such as AOS are treated as adjuncts and left free to be deployed in large numbers, even though they might play a key role in a future ABM system, it would seem only to invite increased breakout anxiety if similar systems were to show up on the Soviet side.

The complexities of the component and development issues may be further illustrated with an example from laser technology. If a ground-based, high-energy laser at the U.S. White Sands ABM test range were tested against a ballistic missile RV that was made to simulate the characteristics of an ICBM RV, that test would be "in an ABM mode'" but also clearly within the permissible bounds of the Treaty.[112] Indeed, using Nitze's development criteria, such a test, even if "in an ABM mode," might fall entirely outside the writ of the Treaty if the laser in question were a pre-prototype collection of plumbing and optics (which is a fair description of the MIRACL laser used in recent lethality tests).[113] This prompts two policy questions: What if the plumbing were collected other than at an ABM test site? What if, in addition, the collection were labelled in Cyrillic?

Let us look at the issue from another perspective. Would such a test still be within bounds of the Treaty if the same laser were to point its beam at an orbiting relay mirror? Assume that the mirror had the ability (via onboard pointing and tracking) to redirect the beam to another mirror or to a target, with lethal effect. Some of the equipment involved would be firmly anchored to the earth, but crucial parts would be based in orbit. The laser alone might be able to intercept RVs coming toward it; augmented by the mirrors, it might be able to strike at boosters rising half a world away. Should the relay mirror that permits that capability extension be regulated as a component, or should it be treated as an unregulated "adjunct?" The mirror(s) would serve functions that are part sensor, part launcher, and part interceptor, and would extend the capabilities of the laser. Should development and testing of such mirrors be permissible without the large laser? What if the mirrors and their pointing and tracking systems were tested in orbit, with a companion orbital laser that was several orders of magnitude less powerful than that required for BMD? What if it were only tested against satellite targets? Under Nitze's criteria, orbiting targets are not considered to have the flight characteristics of a strategic ballistic missile, and thus space targets are fair game. This relationship between BMD and anti-satellite weapons is sufficiently important to merit further consideration.

BMD and Anti-satellite Weapons. ASAT and BMD are conceptually distinct missions that cannot be completely separated politically or technically.[114] Constraints on ASAT systems are imposed indirectly by the Outer Space Treaty of 1967 (which prohibits the placing in orbit of nuclear or other weapons of mass destruction) and by the Limited Test Ban Treaty of 1963 (which prohibits the testing of all nuclear weapons, except underground). The ABM Treaty also imposes some limitations on ASATs: They may not be tested in an ABM mode or be given capabilities to counter strategic ballistic missiles in flight trajectory. Otherwise there are no international limits on the development, testing, or deployment of anti-satellite weapon systems, regardless of their basing mode. However, experience gained by the development and testing of space-based kinetic or directed energy ASATs could prove valuable to the design and development of a space-based BMD, even though ASAT systems may not need the power of their larger cousins. If an ASAT system were never tested in an ABM mode, its ABM capability could be difficult to prove; such a system might not be of any great military significance as an ABM. But technology having potential future ABM application could be developed under the aegis of an ASAT program, or a space-based technology developed under SDI could be tested against space objects "in an ASAT mode," thereby furthering capabilities that the Treaty seeks to ban. In its 1986 report to Congress, SDIO argues that "intercepts of certain orbital targets simulating anti-satellite weapons can clearly be compatible" with the Treaty, so long as the devices involved are not capable of substituting for an ABM component and are not tested in an ABM mode. On September 5, 1986, the SDI program did successfully test a sensor and interceptor, launched into space from Cape Canaveral by a Delta rocket, against a satellite target. The experiment, known as Delta 180 or *Vector Sum,* also measured infrared signatures from a sounding rocket launched from the White Sands Missile Range.[115]

In conducting this test, SDI demonstrated the limited impact of Congressional efforts to constrain U.S. ASAT testing via program-specific, as opposed to action-specific limitations. Since 1984, Congress has restricted tests of the U.S. Space Defense System against objects in space, moving from a test quota to an outright ban in fiscal 1986. By these Congressional actions, the United States responded to the ASAT test moratorium declared by then-General Secretary Andropov in August 1983. The Congressional ban applied only to one missile system, launched at low-earth-orbit targets by an Air Force F-15 jet; it did not cover testing of other systems against space objects, nor did it ban so-called "point in space" tests, where the missile's sensor locks onto the heat of a bright star. There is no indication that the *Vector Sum* experiment was intended to circumvent the Congressional ASAT ban. Rather, without even trying, it demonstrated a U.S. co-orbital ASAT capability similar to that possessed by the Soviet Union, testing of which had been suspended.[116]

Laser ASAT programs could demonstrate the feasibility and practicality of space-based and space-directed lasers for BMD applications. Dean Wilkening calculates that a laser based in low earth orbit and tested against targets in geosynchronous orbits (not less than 35,000 kilometers distant if the laser is orbited at 1,000 kilometers altitude) would require pointing and tracking accuracy and beam brightness equivalent to that required for the BMD mission at one-tenth that range, a commonly considered range for boost-phase BMD.[117]

Finally, long-range, ground-based BMD interceptors deployed as permitted by the Treaty could play a limited but effective role as ASAT interceptors, especially if they were not nuclear armed (greatly increasing their utility in scenarios short of nuclear war). In July 1984, the United States staged a successful, non-nuclear intercept of an RV launched by a Minuteman ICBM from the Western Missile Test Range at Vandenberg Air Force Base, California. The Homing Overlay Experiment (HOE), launched from the U.S. ABM test range at Kwajalein Atoll (using a modified Minuteman booster), maneuvered the spokes of an infrared-guided steel "wagon wheel" into the RV's path. The force of the head-on collision at several kilometers per second disintegrated both interceptor and target well above the earth's sensible atmosphere; but the RV could just as well have been a satellite in low earth orbit. The HOE program has been superseded by the Exoatmospheric Reentry-vehicle Interceptor Subsystem (ERIS), a somewhat more compact hit-to-kill experiment. SDIO will test ERIS with only one kill vehicle, as prescribed by the Treaty.[118] However, a multi-kill-vehicle, ERIS-like experiment theoretically could be conducted against satellite targets as part of an ASAT development program. Although in operational situations the ASAT mission would the easier of the two (since the path of a target satellite would be known well in advance), the kinematics of low-earth-orbit satellites and ICBM RVs pose similar challenges to a long-range interceptor. Tests of such a "MIRVed" system against an object in orbit could add materially to the development of MIRVed ABM interceptors, which is prohibited by the Treaty.

NOTES

1. For an assessment of the naval agreements, see Hedley Bull, "Strategic Arms Limitation: The Precedent of the Washington and London Naval Treaties," in Morton A. Kaplan, ed., *SALT: Problems and Prospects* (Morristown, N.J.: General Learning Press, 1973), pp. 26–52.

2. Raymond Garthoff, *Detente and Confrontation* (Washington, D.C.: Brookings Institution, 1985), pp. 42 ff.

3. Michael Krepon, "Decontrolling the Arms Race," *Arms Control Today* (March/April 1984): 2.

4. General Advisory Committee on Arms Control and Disarmament, "A Quarter Century of Soviet Compliance Practices Under Arms Control Agreements, 1958–1983, Summary" (Washington, D.C.: Oct. 1984, Mimeographed). For counterpoint, see Jonathan Rich, "The General Advisory Committee Report: An FAS Staff Study" (Washington, D.C.: Fed-

eration of American Scientists, n.d., Mimeographed). General Richard Ellis, former head of SAC and current U.S. Commissioner to the SALT Standing Consultative Commission, and Lieutenant General John Chain, former Director of the Bureau of Political/Military Affairs in the Department of State and current head of SAC, testified before Congress in closed session that Soviet activities generally are compliant with the terms of arms control agreements. See Walter Andrews, "Soviets Said to Abide by Most Arms Terms," *Washington Times,* 9 Jan. 1986, p. 4.

Something approaching an absolute standard of verification was reflected in U.S. Defense Department efforts to have the Reagan administration's 1987 report, "Soviet Noncompliance," state that the Soviet Union "is preparing" a prohibited ABM defense of its national territory; the old wording ("may be preparing") was retained in the final report, more consistent with the tenuous nature of the evidence presented. Michael Gordon, "U.S. Is Debating Role of Three New Soviet Radars," *NYT,* 19 Dec. 1986, p. A11; White House, Office of the Press Secretary, *The President's Unclassified Report on Soviet Noncompliance with Arms Control Agreements* (Washington, D.C.: U.S. Government Printing Office, 10 Mar. 1987). For counterpoint, see James Rubin and Matthew Bunn, "Analysis of the President's Report on Soviet Noncompliance" (Washington, D.C.: Arms Control Association, 12 Mar. 1987, Mimeographed).

5. Stephen M. Meyer, "Verification and Risk in Arms Control," *International Security* (Spring 1984): 118.

6. Ibid., p. 124.

7. Ibid., p. 113.

8. Ibid., p. 118.

9. William J. Durch, "Antisatellite Weapons, Arms Control Options, and the Military Use of Space," Report No. AC3PC103 for the U.S. Arms Control and Disarmament Agency. (Cambridge, Mass.: Harvard University, Center for Science and International Affairs, November 1984), pp. 34–35.

10. Meyer, "Verification and Risk," note 5, p. 115–17. Of course it may be the objective of policy to so undermine the Treaty. The side pursuing such a policy will soon lose credibility unless it can control internal dissent, making this a non-viable long-term policy for a pluralistic government.

11. The latter elements are drawn from the U.S. proposal on verifying reductions of intermediate-range nuclear forces to 100 missiles apiece, submitted in the Geneva talks in March 1987. John H. Cushman, "U.S. is Demanding Wide Verification for a Missile Pact," *NYT,* 13 Mar. 1987, pp. 1, 12.

12. Meyer, "Verification and Risk," note 5, p. 124.

13. Herbert Lin, "New Weapon Technologies and the ABM Treaty." (Cambridge, Mass.: Defense and Arms Control Studies Program, Center for

International Studies, MIT, 1 June 1987, Mimeographed), pp. 85–86. The brightness range given assumes lasers of 25 to 100 watts power, apertures of 0.25 to 1.0 meters, and operating wavelengths from microwave to visible. Aperture limits, based on the maximum power handling ability of mirrors of a given size, might be a useful supplement to prevent circumvention via low-power tests of large-aperture lasers. A perfectly reflective 1-meter mirror might be able to handle up to 20 to 30 megawatts of power. For a discussion of mirror power handling limits, see Dean A. Wilkening, "Space-Based Weapons," in William Durch, ed., *National Interests and the Military Use of Space* (Cambridge, Mass.: Ballinger Publishing Company, 1984), pp. 145–50.

14.   John Pike, "Limitations on Space Weapons, A Preliminary Assessment" (Washington, D.C.: Federation of American Scientists, February 1987, Mimeographed), p. 29.

15.   Raymond Garthoff, "SALT I: An Evaluation," *World Politics* (Spring 1978): 17; and testimony of Albert Carnesale, former member of the U.S. SALT I delegation, in U.S. Congress, Senate, Armed Services Committee, *Department of Defense Authorization . . . Fiscal Year 1986, Part 7: Strategic and Theater Nuclear Forces*, 99th Cong., 1st sess., S. Hearing 99–58 pt. 7, 18 Mar. 1985, p. 4020.

16.   Personal communication of George Rathjens with the author, gratefully acknowledged.

17.   Blechman and Utgoff estimate the ten-year systems costs of several notional U.S. strategic defense systems, incorporating air as well as missile defenses and using the best of near-term technologies. Air defenses account for two-thirds of the $160 billion cost (in 1987 dollars) of defending strategic forces. Their territorial defense variants, which incorporate the force defenses, cost $670 to $770 billion over ten years. The assumptions that Blechman and Utgoff use to size defensive forces are highly favorable to the defense (they give a complete list of assumptions on p. 102). For example, all intercept/kill mechanisms work with a 90 percent single-shot probability of kill (SSPK); the system acquisition costs tail off according to a 90 percent learning curve (with each doubling of production, unit costs drop 10 percent); and there is no responsive Soviet ASAT challenge to space-based components. Changing the assumed SSPK to 80 percent (still ahistorically high for air defense effectiveness) increases overall system costs by 20–30 percent (p. 103). Even at 90 percent SSPK, defending strategic forces would cost roughly as much as "27 wings of F-15s or 14 armored divisions" (p. 134). The territorial defense variants would cost about as much per year as "the Navy's or the Air Force's total investment budget in fiscal 1986, and about twice as much as the Army invested that year." We would, in short, be buying the equivalent of another military service. The Soviet Union has had such a separate service for

strategic defense, the PVO Strany (now Voiska PVO), for more than thirty years. Barry M. Blechman and Victor A. Utgoff, *Fiscal and Economic Implications of Strategic Defense*, SAIS Papers in International Affairs Number 12 (Boulder, Colo.: Westview Press, 1986).

18. John Newhouse, *Cold Dawn: The Story of SALT* (New York: Holt, Rinehart & Winston, 1973), p. 73; Lawrence Freedman, *Evolution of Nuclear Strategy* (New York: St. Martin's Press, 1982), pp. 345, 352; and Ted Greenwood, *Making the MIRV* (Cambridge, Mass.: Ballinger Publishing Company, 1975), esp. p. 165.

19. For excerpts, see Benson Adams, *Ballistic Missile Defense* (New York: American Elsevier, 1971), p. 166.

20. *Izvestiia*, 24 Aug. 1972 and *Pravda*, 30 Sept. 1972, as cited by Raymond Garthoff, "SALT and the Soviet Military," in Jiri Valenta and William Potter, eds., *Soviet Decisionmaking for National Security* (London: George Allen & Unwin, 1984), p. 140.

Colin Gray denies that the Soviet Union could have afforded a larger offensive program than that actually deployed in the 1970s, and suggests that if a major BMD program had been part of the modernization effort, the resulting growth in offensive capability "might even have been smaller." See Colin Gray, "A New Debate on Ballistic Missile Defense," *Survival* (March/April 1981): 63. However, considering the excess throwweight available on the SS-18 ICBM, and the availability of 6-warhead SS-19s to substitute for less accurate, 4-warhead SS-17s, it would seem that at least another 2,000 RVs could have been added to a non-SALT-constrained Soviet ICBM force with not all that great an effort. And funds to sustain a BMD program would need not have been drawn from the ICBM account.

21. U.S. Congress, Senate, Committee on Foreign Relations, *Hearings on the SALT I Agreements*, 92nd Cong., 2d sess., 19 June 1972, p. 399.

22. According to Iklé, "We now find that the intellectual foundation for the ABM Treaty and of SALT has been washed away" by strategic modernization. Remarks before the U.S. House Republican Study Committee; see *Washington Times*, 10 June 1985, p. 4; see also Colin Gray, "Moscow is Cheating," *Foreign Policy* (Fall 1984): 148–50; and U.S. Department of Defense, Office of the Secretary Defense, *Fiscal Year 1987 Report of the Secretary of Defense to the Congress*, p. 82.

23. For a discussion of offense survivability, crisis stability, and defense, see Charles Glaser, "Why Even Good Defenses May Be Bad," *International Security* (Fall 1984): 108; and Dean A. Wilkening, et al., "Strategic Defenses and First-Strike Stability," *Survival* (March/April 1987): esp. 161.

24. Absence of BMD helps the bombers because relatively abundant, lower accuracy SLBM warheads can be used to clear reliable bomber penetration corridors through Soviet air defenses. The United States did begin to

search for alternative ICBM basing modes in the latter 1970s, but the weapon of choice (the MX) was too heavy to be off-road mobile or to use ordinary U.S. highways. Ten years later (and then only under Congressional pressure), the United States began to develop a small mobile missile. By then, the Soviet Union was already starting to flight test its own small mobile missile (the SS-25). For an assessment of MX issues, see U.S., Congress, Office of Technology Assessment, *MX Missile Basing* (Washington, D.C.: U.S. Government Printing Office, 1981). On the newer mobiles, see William Durch and Peter Almquist, "East-West Military Balance," in Blechman and Luttwak, eds., *International Security Yearbook, 1984-85.* (Boulder, Colo.: Westview Press, 1985), pp. 14-15, 21-22, 31-33.

25. See, for example, Gray, "A New Debate," note 20, p. 64.

26. The White House, *The President's Strategic Defense Initiative,* January 1985, p. 3.

27. There is some question as to whether SLBMs should be considered "prompt" retaliatory systems. The question arises not from SLBM performance once launched (their flight times are somewhat shorter than ICBMs', Soviet warning of SLBM attack is much shorter [since the Soviet Union lacks satellite early warning of SLBM launch], and D5 will have MX-like accuracy and firepower), but from evident uncertainties of command and control that may delay prompt transmission of emergency war orders or action messages to an SSBN on patrol. For discussion, see Ashton Carter, "Communications Technologies and Vulnerabilities," and Stephen M. Meyer, "Soviet Nuclear Operations," in Carter, Steinbruner, and Zraket, eds., *Managing Nuclear Operations*,(Washington, D.C.: Brookings Institution, 1987), pp. 237, 479; Theodore A. Postol, "The Trident and Strategic Stability," *Oceanus* (Summer 1985): 47-48.

28. Charles A. Zraket, "Strategic Defense: A Systems Perspective," *Daedalus* (Spring 1985): 114.

29. For other discussions of uncertainty and BMD, see Charles Glaser, "Do We Want the Missile Defenses We Can Build?" *International Security* (Summer 1985); John Wright, "Uncertainty and the Strategic Defense Initiative," *Nature* (23 January 1986): 275-79; and Lawrence Freedman, "The 'Star Wars' Debate: The Western Alliance and Strategic Defense: Part II," *Adelphi Paper 199* (London: IISS, 1985), p. 41.

30. Christoph Bertram, "Strategic Defense and the Western Alliance," *Daedalus* (Summer 1985): 279.

31. Nuclear policy related disputes within NATO have included those over the Thor and Jupiter Missiles (1958-59), the Skybolt missile (1962), the Multilateral Force (1963), the adoption of MC14/3 (the doctrine of flexible response [1967]), enhanced radiation weapon (neutron bomb) deployment (1977-78), the impact of SALT on extended deterrence (1974-1979), and (partly in consequence) modernization of intermediate-range

nuclear forces (the Pershing II and ground-launched cruise missile decision [1979-1983]). See Ivo H. Daalder and Lynn Page Whittaker, "SDI's Implications for Europe," in Stephen Flanagan and Fen Hampson, eds., *Securing Europe's Future* (London: Croom Helm Press, 1986), pp. 44, 56.

Doubts about the adequacy of NATO's conventional capabilities and concerns about the cost of primary reliance on conventional defenses have formed the backdrop for all these nuclear debates. For pessimistic assessments of NATO's current conventional capabilities, see U.S., Congress, Senate, Armed Services Committee, Report of Senators Nunn and Bartlett, *NATO and the New Soviet Threat*, Committee Print, 95th Cong., 1st sess., (January 1977); and U.S. Congress, Congressional Budget Office, *U.S. Ground Forces: Design and Cost Alternatives for NATO and Non-NATO Contingencies* (Washington, D.C.: U.S. Government Printing Office, Dec. 1980). For cautiously optimistic assessments of the current balance, see John Mearsheimer, *Conventional Deterrence* (Ithaca, N.Y.: Cornell University Press, 1983); Barry R. Posen, "Measuring the European Conventional Balance: Coping with Complexity in Threat Assessment," *International Security* (Winter 1984-85): 47-88; and Joshua M. Epstein, *The 1987 Defense Budget* (Washington, D.C.: Brookings Institution, 1986), esp. pp. 28 ff.

32.   Ivo H. Daalder, *The SDI Challenge to Europe* (Cambridge, Mass.: Ballinger Publishing Company, 1987), p. 40.

33.   Bertram, "Strategic Defense," note 30, p. 294.

34.   Jacquelyn K. Davis and Robert L. Pfaltzgraff, Jr., "Strategic Defense and Extended Deterrence: A New Transatlantic Debate," *National Security Paper No. 4* (Cambridge, Mass.: Institute for Foreign Policy Analysis, February 1986), pp. 14-15. The authors take a view of Soviet BMD capabilities that is much less sanguine than even the Reagan administration's.

35.   Ibid., p. 17. David Yost expresses similar concerns. See David Yost, "Soviet Ballistic Missile Defense and NATO," *Orbis* (Summer 1985): 281-91. Both works focus on the danger of Soviet "breakout" of the Treaty.

36.   Sidney Drell, Philip Farley, and David Holloway, *The Reagan Strategic Defense Initiative: A Technical, Political and Arms Control Assessment* (Cambridge, Mass.: Ballinger Publishing Company, 1985), p. 76. For an assessment that reaches much the same conclusions about European views and concerns but is more sanguine about solutions to them, see David S. Yost, "European Anxieties About BMD," *Washington Quarterly* (Winter 1984-85). For a cost assessment of deterrence enhancing requirements for NATO that could follow from deployment of very good U.S.-Soviet strategic defenses, see Blechman and Utgoff, note 17, pp. 64-66.

37.   Daalder and Whittaker, note 31, pp. 57-58.

38.   The text of the Treaty and associated statements used in this overview may be found in U.S. Arms Control and Disarmament Agency, *Arms Con-*

trol and Disarmament Agreements, 1980 Edition (Washington, D.C.: U.S. Government Printing Office, 1980). This edition corrected some discrepancies in the labelling of several Agreed Statements in earlier editions.

On April 7, 1972 the U.S. SALT Delegation stated that the United States

> would consider a launcher, missile, or radar to be "tested in an ABM mode" if any of the following events occur: (1) a launcher is used to launch an ABM interceptor missile; (2) an interceptor missile is flight tested against a target vehicle which has a flight trajectory with the characteristics of a strategic ballistic missile trajectory, or is flight tested in conjunction with the test of an ABM interceptor missile or an ABM radar at the same test range, or is flight tested to an altitude inconsistent with the interception of targets against which air defenses are deployed, (3) a radar makes measurements on a cooperative target vehicle of the kind referred to in item (2) above during the reentry portion of its trajectory or makes measurements in conjunction with the test of an ABM interceptor missile or an ABM radar at the same test range. Radars used for purposes such as range safety or instrumentation would be exempt from application of these criteria.

Ibid., p. 147. The SDIO released unclassified synopses of classified Agreed Statements on "testing in an ABM mode" worked out in 1978 and 1985:

> An interceptor missile is considered to be "tested in an ABM mode" if it has attempted to intercept (successfully or not) a strategic ballistic missile or its elements in flight trajectory. Likewise a radar is considered to be "tested in an ABM mode" if it performs certain functions such as tracking and guiding an ABM interceptor missile or tracking strategic ballistic missiles or their elements in flight trajectory in conjunction with an ABM radar which is tracking and guiding an ABM interceptor missile. "Strategic ballistic missiles . . ." include ballistic target-missiles with the flight trajectory characteristics of strategic ballistic missiles over the portions of the flight trajectory involved in testing.

U.S., Department of Defense, Strategic Defense Initiative Organization, *Report to the Congress on the Strategic Defense Initiative* (Washington, D.C.: U.S. Government Printing Office, June 1986), p. C7. (Hereafter cited as *1986 SDIO Report.*)

39. Malcolm Russell, "Soviet Legal Views on Military Space Activities," in Durch, *National Interests*, note 13, pp. 211-12.

40. Abraham Chayes, "An Inquiry into the Workings of Arms Control Agreements," *Harvard Law Review* 85:5 (March 1972): 957-59; John Rhinelander, "The SALT Agreements," in Rhinelander and Willrich, eds., *SALT: The Moscow Agreements and Beyond* (Glencoe, N.Y.: The Free Press, 1974), p. 142; and *Vienna Convention on the Law of Treaties*, Article 60, 63 AJIL 875 (1969), cited in William W. Bishop, Jr., *International Law: Cases and Materials*, 3d ed. (Boston: Little, Brown, 1971), pp. 205-206.

41. Gerard Smith, *Doubletalk: The Story of the First Strategic Arms Limitation Talks* (Garden City, N.Y.: Doubleday, 1980), pp. 301 ff.

42. For excellent maps of U.S. and Soviet radar coverage, see Thomas Longstreth, John Pike, and John Rhinelander, *The Impact of U.S. and Soviet*

Ballistic Missile Defense Programs on the ABM Treaty (Report prepared for the National Campaign to Save the ABM Treaty, Washington, D.C., March 1985), pp. 69–74.

43. Michael Gordon, "Defense Department Rebuffed on Soviet ABM Threat," *NYT*, 5 Mar. 1987, p. A10; and U.S., Department of Defense, *Soviet Military Power, 1987* (Washington, D.C.: U.S. Government Printing Office, annual), pp. 47–49. (Hereafter cited as *SMP* [year].)

44. Michael Gordon, "Soviets Finishing Large Radar Center in Siberia," *NYT*, 23 Nov. 1986, p. 8.

45. *SMP, 1985*, pp. 45–46.

46. Leslie Gelb, "U.S. Is Challenged on Soviet Radar," *NYT*, 15 Mar. 1985, p. 12.

47. R. Jeffrey Smith, "Administration at Odds over Soviet Cheating," *Science* (March 22, 1985): 1443; Adams, *Ballistic Missile Defense,* note 19, p. 145; Gordon, "Soviets Finishing Large Radar," note 44; John Toomay, "Warning and Assessment Sensors," in Carter, et al., *Managing Nuclear Operations,* note 27, p. 299.

    Charles Gellner of the Congressional Research Service argues that Krasnoyarsk will not be in violation of the Treaty until it is actually operational (perhaps in 1988). However, this argument would permit both sides to partially build any number of ABM components legally. See Charles Gellner, "Violations of Arms Control Agreements: Validity of Some Salient Arguments." (Paper presented to the 1986 Annual Meeting of the Section on Military Studies, International Studies Association, Cambridge, Mass., December 1986).

48. U.S., Congress, Senate, Committee on Foreign Relations, *Briefing on SALT I Compliance,* 96th Cong., 1st sess., 25 Sept. 1979, p. 50.

49. See Rubin and Bunn, "Analysis of the President's Report," note 4, pp. 2–4; and David C. Morrison, "Radar Diplomacy," *National Journal* (3 January 1987), pp. 17–18.

50. Ibid.

51. *SMP*, 1987, note 43, p. 49.

52. U.S. Senate, *Briefing on SALT I Compliance,* note 48, p. 26.

53. Ibid., p. 53.

54. U.S. Arms Control and Disarmament Agency, *Soviet Noncompliance 1986* (Washington, D.C.: U.S. Government Printing Office, February 1986), p. 4.

55. Rhinelander, "The SALT Agreements," note 40, pp. 134–35.

56. R. Jeffrey Smith, "Soviet Radars of Concern to U.S. Removed," *Washington Post,* 25 Feb. 1987, p. 20.

57. George Schneiter, "The ABM Treaty Today," in Ashton Carter and David Schwartz, eds., *Ballistic Missile Defense* (Washington, D.C.: Brookings Institution, 1984), p. 226n.

58. U.S. Congress, Senate, *Soviet Strategic Force Developments, Hearings Before a Joint Session of the Subcommittee on Strategic and Theater Nuclear Forces, Senate Armed Services Committee, and the Defense Subcommittee, Senate Committee on Appropriations*, 99th Cong., 1st sess., S. Hearing 99–335, testimony of Robert M. Gates and Lawrence K. Gershwin, 26 June 1985, pp. 5–6.

59. *Aviation Week and Space Technology* (16 Jan. 1984): 16.

60. Gordon, "Defense Rebuffed," note 43; Longstreth, et al., *Impact*, note 42, p. 57. Note that photo reconnaissance coverage is discontinuous, that ninety minutes is the minimum revisit period if only one satellite is available, and that immediate revisiting is possible only with extensive maneuvering to compensate for the earth's rotation under the orbiting satellite. Without maneuvering, the average revisit period in daylight is about 3.5 days. See Jeffrey Richelson, "Technical Collection and Arms Control," in William C. Potter, ed., *Verification and Arms Control* (Lexington, Mass.: Lexington Books, 1985), p. 175.

61. *Statement of Secretary of Defense Melvin Laird Before the House Armed Services Committee on the Fiscal 1971 Defense Program and Budget*, 2 Mar. 1970, p. A15. Declassified 25 June 1975. DDRS 1975, 153A.

62. Smith, *Doubletalk*, note 41, p. 32.

63. U.S. Senate, *Briefing on SALT I Compliance*, note 48, pp. 49–50; *1986 SDIO Report*, note 38, p. C7.

64. The U.S. Defense Department now identifies two variants of the SA-12 interceptor: the tactical A variant (NATO code name "GLADIATOR") and the longer range B variant (code name "GIANT"), which is still experimental. GIANT apparently has been tested against tactical ballistic missile targets, reportedly including the 900 kilometer range SS-12 and the 2,000 kilometer range SS-4, scoring one intercept in twenty attempts. Gordon, "Defense Rebuffed," note 43; *SMP, 1987*, p. 61; Longstreth, et al., *Impact*, note 42.

65. See Benoit Morel and Theodore A. Postol, "A Technical Assessment of Potential Threats to NATO from Non-Nuclear Soviet Tactical Ballistic Missiles" and idem, "Anti-Tactical Ballistic Missile Technologies and NATO," in Jeffrey Boutwell and Donald Hafner, eds., *European Missile Defenses: ATBMs and Western Security* (Cambridge, Mass.: Ballinger Publishing Company, forthcoming). See also Benoit Morel, "ATBM—A Solution in Search of a Problem," *Bulletin of the Atomic Scientists* (May 1987): 39–41.

66. Herbert Lin, "New Weapon Technologies," note 13, pp. 53 ff.

67. Morel and Postol, "Anti-Tactical Ballistic Missile Systems," note 65.

68. See U.S., Congress, Senate, Armed Services Committee, *DOD Authorization for FY 1986*, 99th Cong., 1st sess., S. Rept. Hearing 99–58, pt. 7, 26 Feb. 1985, p. 3577.

69. See, for example, U.S., Congress, Office of Technology Assessment, *Ballistic Missile Defense Technologies* (Washington, D.C.: OTA, 1985); Harold Brown, "Is SDI Technically Feasible?" *Foreign Affairs* (Winter 1985–86): 435–54; and *Report to the American Physical Society of the Study Group on the Science and Technology of Directed Energy Weapons.* Reprinted in *Reviews of Modern Physics* (July 1987). (Hereafter, *APS Report.*)

70. U.S., Department of Defense, Strategic Defense Initiative Organization, *Report to the Congress on the Strategic Defense Initiative, 1985*, Appendix B (Washington, D.C.: U.S. Government Printing Office, April 1985), pp. B4–B9. Monitoring Defense Department compliance with Treaty development and testing constraints is the responsibility of the office of the Undersecretary of Defense for Research and Engineering (USDRE) as assigned by a Defense Department Directive of January 1973. USDRE interprets Treaty obligations on a case-by-case basis, in consultation with the Defense Department General Counsel. There is no formal interagency process or forum for monitoring U.S. compliance with the Treaty, although ACDA does not accept exclusive USDRE jurisdiction in Treaty compliance interpretation. Ibid., p. B3.

71. Oberdorfer, "ABM Reinterpretation: A Quick Study," *Washington Post*, 22 Oct. 1985. The administration's own legal experts with extensive experience in arms control were not consulted in crafting the new interpretation, and the rebuttals to the new interpretation that they drafted were ignored by the administration's political appointees. Michael Gordon, "Ex-Aide Says Reagan Got Flawed Advice on ABMs," *NYT*, 30 Apr. 1987, p. A3.

72. Ibid.; see also Don Oberdorfer, "Top-Level Fight Led to ABM Policy Shift," *Washington Post*, 17 Oct. 1985; Walter Pincus, "New U.S. Arms Stance Reported," *Washington Post*, 23 Oct. 1985.

73. Ibid.

74. Abraham D. Sofaer, "The ABM Treaty and the Strategic Defense Initiative," *Harvard Law Review* 99:8 (June 1986): 1972–85.

75. Note, however, that Secretary Rogers' report to the president on the SALT agreements, which was included in the Nixon administration's transmittal of the ABM Treaty and Interim Agreement to the Senate in June 1972, stated in the section on "Future ABM Systems:" "Article II(1) defined an ABM system in terms of its *function* as 'a system to counter strategic ballistic missiles or their elements in flight trajectory,' noting that such systems 'currently' consist of ABM interceptor missiles, ABM launchers and ABM radars. [Emphasis added.]" U.S., Arms Control and Disarmament Agency, *Documents on Disarmament 1972* (Washington, D.C.: U.S. Government Printing Office, 1972), pp. 272–73. See also Abraham Chayes and Antonia Chayes, "Testing and Development of

'Exotic' Systems Under the ABM Treaty: The Great Reinterpretation Caper," *Harvard Law Review* 99:8 (June 1986): 1956-71.

76. Sofaer, "The ABM Treaty," note 74, pp. 1974-75, 1979. Sofaer notes (on p. 1979) that his assessment is confirmed by Nitze. However, it was not Nitze but Garthoff who negotiated Agreed Statement D. See Raymond Garthoff, "Correspondence: On Negotiating with the Russians," *International Security* (Summer 1977): 107-109.

77. Sofaer, "The ABM Treaty," note 74, pp. 1980-84.

78. Oberdorfer, "Top-Level Fight," note 72.

79. Charles Mohr, "'Option' Sought to Deploy Space Shield Soon," *NYT*, 26 Mar. 1986, p. A21.

80. *Pravda* account quoted by Philip Taubman, "Soviet Military Chief Accuses U.S. of Distorting Terms of ABM Treaty," *NYT*, 19 Oct. 1985; Leslie Gelb, "Reagan Reported to Stay Insistent on 'Star Wars' Test," *NYT*, 24 July 1986, p. 1.

81. Ibid.; "Excerpts from Speech by Gorbachev About Iceland Meeting," *NYT*, 15 Oct. 1986, p. A12.

82. White House, Office of the Press Secretary, *Press Briefing by Admiral John M. Poindexter, National Security Advisor*. (Washington, D.C.: Mimeographed.) The U.S. proposal tabled at Reykjavik pledged "strictly to observe" Treaty provisions for a period of ten years "while continuing research, development and testing, which are permitted by the ABM Treaty. ... At the end of the ten-year period, either side could deploy defenses if it so chose unless the parties agree otherwise." *NYT*, 18 Oct. 1986, p. 5. The particular construction used in the U.S. proposal ("research, development and testing, which are permitted ...") implies that all activities in these categories are permitted under the Treaty. A limiting construction ("research, development and testing that are permitted ...") acknowledges that the Treaty imposes constraints in these areas.

83. *APS Report*, note 69.

84. Colin Norman, "Debate Over SDI Enters New Phase," *Science* (16 Jan. 1987): 277-80, reports that Edward Teller, Alexander Haig, and other "conservative legislators and supporters of SDI" wrote to Reagan in October 1986 as follows: "We are deeply concerned that a SDI research program which has no definite consequences for defense of America and its allies within the next 10 years will not be politically sustainable" (p. 277).

85. Michael Gordon, "Star Wars Timetable: What Effect at Geneva Talks?" *NYT*, 21 Jan. 1987, p. A14. Tentative plans called for 250 battle satellites with 10 interceptor rockets each, backed by 100 ground-based terminal defense interceptors. William Beecher, "Reagan May Get SDI Plan in March," *Boston Globe*, 24 Jan. 1987, p. 1. For a critique of High Frontier, see the response by the Department of the Army to questions from the Defense Appropriations Subcommittee, U.S., Congress, Senate, Armed

Services Committee, *Hearings on Fiscal 1984 DOD Appropriations, Part 2*, 98th Cong., 1st sess., 1983, pp. 154–55; see also Ashton Carter, *Directed Energy Missile Defense in Space* (Washington, D.C.: Office of Technology Assessment, April 1984), pp. 34–35. Most critics of the concept believe that it can be quickly outstripped by offensive countermeasures.

86.   John Cushman, "Weinberger Gives Strategy Outline on Missile Shield," *NYT*, 13 Jan. 1987, p. 1; Michael Gordon, "Reagan and Advisers Meet on Deploying of Star Wars," *NYT*, 4 Feb. 1987, p. 2; idem, "U.S. Aides Expect Delay of Decision Over ABM Treaty," *NYT*, 8 Feb. 1987, p. 1.

87.   Michael Gordon, "Arms Negotiators Plan New Effort," *NYT*, 1 Feb. 1987, p. 1; William Beecher, "A Clash on Arms Control," *Boston Globe*, 6 Feb. 1987, p. 13; Michael Gordon, "Reagan Reported to Limit Debate at Geneva Talks," *NYT*, 22 Feb. 1987, p. 1; Helen Dewar and R. Jeffrey Smith, "Administration Warned on SDI Testing," *Washington Post*, 25 Feb. 1987, p. 18; Elizabeth Pond, "Broad View of ABM Was Not Proposed," *Christian Science Monitor*, 27 Feb. 1987, p. 12.

88.   West Germany, in particular, urged the United States to adhere to the strict interpretation. Former SALT negotiators rejecting the broad interpretation include Gerard Smith, head of the delegation; General Royal Allison, JCS representative; Albert Carnesale, advisor and ACDA representative; Sidney Graybeal and Raymond Garthoff, delegation working group chairmen; and John Rhinelander, legal advisor to the delegation. The former Defense secretaries (Robert McNamara, Clark Clifford, Melvin Laird, Elliot Richardson, James Schlesinger, and Harold Brown, the latter also an advisor on the SALT I delegation) signed a joint statement sent to Reagan, Cabinet officials, and leaders of Congress. See Michael Gordon, "Allies Surprised by Plans to Speed Star Wars Tests," *NYT*, 6 Feb. 1987, p. 1; idem, "Arms Debate Now Centers on ABM Pact," *NYT*, 17 Feb. 1987, p. A1; idem, "Ex-Defense Secretaries Back Strict View of '72 ABM Pact," *NYT*, 10 Mar. 1987, p. A12.

89.   Colin Norman, "Senator Blasts Administration's Reinterpretation of ABM Treaty," *Science* (19 December 1986): 1489.

90.   Senator Sam Nunn, "Interpretation of the ABM Treaty, Part One: The Senate Ratification Proceedings," Washington, D.C., 11 Mar. 1987, pp. 54, 56, 60, 63; idem, Part Two: Subsequent Practice Under the ABM Treaty, 12 Mar. 1987, esp. pp. 6–7, 22; and idem, Part Three: The Treaty Negotiating Record, 13 Mar. 1987, p. 7. See also U.S. Congress, *Congressional Record*, 100th Cong., 1st sess., 11–13 Mar. 1987, pp. S2967–S2986, S3090–S3095, and S3171–S3173.

Nunn observes that the negotiating record, under international law and practice, ranks third after the plain meaning of the text of an agreement and the actual practice of the parties as a source for interpreting the meaning of a Treaty. See also Norman, note 89, for the views of Senator Levin. Levin expressed his views in a long letter to Secretary of State Shultz, released by his office.

Sofaer confirmed his support of the broad interpretation in a new study sent to the Senate on 30 April 1987, reiterating that ratification testimony appears ambiguous and repeating the argument that such testimony does not accurately reflect the negotiating record and thus was not binding on the Soviets. See Michael Gordon, "New State Department Study Backs Administration Stand on ABM Pact," *NYT*, 4 May 1987, p. A14.

91. The Senate Armed Services Committee followed suit, indirectly, by voting limits on SDI testing not consistent with the traditional interpretation. John Cushman, "Senate Panel Votes Against Testing Space Defense," *NYT*, 6 May 1987, p. A9.

92. Newhouse, note 18, pp. 230–31. The Joint Chiefs of Staff opposed NSDM 127, which directed the U.S. SALT delegation to seek a ban on exotic ABMs and, according to Garthoff, Pentagon appeals opposing that instruction delayed negotiations for over a month. See Garthoff, "Correspondence," note 76, p. 108. For an exhaustive review of Treaty ratification testimony completed before SDI and the reinterpretation controversy, see Alan M. Jones, Jr., "Implications of Arms Control Agreements and Negotiations for Space-Based BMD Lasers," in Keith B. Payne, ed., *Laser Weapons in Space, Policy and Doctrine* (Boulder, Colo.: Westview Press, 1982), esp. pp. 37–55. Jones' conclusions are essentially identical to Nunn's.

93. Nunn, *Interpretation,* note 90, Part 3, p. 6.

94. When the Reagan administration announced in mid-October 1985 that the traditional interpretation would remain U.S. policy, a White House spokesman noted that this would be the case "as long as we consistently receive the needed budgetary support" for SDI. The statement was retracted a few hours later, but reappeared in the fiscal 1987 SDI *ACIS*, which gives October 1985 as the date of a presidential determination that adherence to the traditional interpretation of the Treaty would be conditioned upon adequate budgetary support for SDI, on the grounds that "work arounds" are costly and that a reduced SDI budget could not afford both to do them and still keep to its original program milestones. See Walter V. Robinson, "White House Backs Shultz's View on ABM Pact," *Boston Globe*, 18 Oct. 1985, p. 13; U.S., Congress, Senate, Foreign Relations Committee, *Fiscal Year 1987 Arms Control Impact Statements*, 99th Cong., 2nd sess., S. Rept. 99-147, April 1986, p. 50.

95. See Garthoff, "Correspondence," note 76, pp. 107–109; also Gordon, "Arms Debate," note 88. At a March 1986 conference on the future of the Treaty, former members of the U.S. delegation confirmed that Agreed Statement D was intended in part to preserve the right to deploy a fixed,

land-based, exotic ABM: The Defense Department was concerned that the Treaty would place too airtight a lid on exotics. Agreed Statement D reserved the right to later negotiate deployment criteria for an exotic system comparable to those in Article III, which governs deployment of traditional components. Seen in this light, Agreed Statement D is neither redundant nor all-encompassing where exotics are concerned, but a reflection of organizational interest protection. William Durch, "Minutes of ABM Treaty Technological/Legal Issues" (Conference held at the Center for International Studies, MIT, Cambridge, Mass., 7–8 Mar. 1986), author's minutes.

96. On Soviet support for strict interpretation, see note 80. On the SCC, see Robert W. Buchheim and Dan Caldwell, *The U.S.-USSR Standing Consultative Commission: Description and Appraisal,* Working Paper No. 2, Center for Foreign Policy Development (Providence, R.I.: Brown University, May 1983); and R. Jeffrey Smith, "Arms Agreement Breathes New Life into SCC," *Science* (9 August 1985): pp. 535–36.

97. U.S., Congress, House, Foreign Affairs Committee, *ABM Treaty Interpretation Dispute,* 99th Cong., 1st sess., 22 Oct. 1985, p. 38.

98. In mid-1980, U.S. Defense Secretary Harold Brown did redirect U.S. high-energy laser programs from ground-based, near-term applications to longer term space-based applications, in particular for satellite defense. However, at the same time the Carter administration explicitly supported the traditional interpretation of the Treaty in its annual *ACIS*. "U.S. Effort Redirected to High Energy Lasers," *Aviation Week and Space Technology* (July 28, 1980): 50–56; and U.S., Congress, Senate, Foreign Relations Committee, *Fiscal Year 1982 Arms Control Impact Statements,* Joint Committee Print, 97th Cong., 1st sess., February 1981, pp. 195–96.

99. R. Jeffrey Smith, "Legal Hurdles Remain for Key SDI Tests," *Washington Post,* 10 Feb. 1987, p. 11. The interceptors envisioned for the space-based component of the phase one BMD system would carry on-board infrared guidance and destroy their targets in collision. Early 1970s-vintage ABM interceptors were command guided (carried no on-board guidance) and had nuclear warheads. Some analysts argue that these differences mean that the new rockets are based on "other physical principles" (the phrase in Agreed Statement D defining an exotic technology), and thus are testable in space under the broad interpretation. In 1986, however, SDIO Director James Abrahamson testified that ground-based, infrared-guided, non-nuclear interceptors could be deployed under Article III of the Treaty because they are *not* an exotic technology.

100. On the radar question in SALT, see Smith, *Doubletalk,* note 41, chapter 10. On the problems presented by new technologies, see Lin, "New Weapon Technologies," note 13, pp. 13–15.

101. Durch, "Minutes," note 95.

102. In connection with Article V restrictions on mobile components, Sherr contrasts the use of *sozdavat* with the term used to translate "development" in Article IV, which governs activities at ABM test ranges. That term (*razrabotka*) denotes development "in the narrow sense of 'working something out.'" Alan Sherr, "The Languages of Arms Control," *Bulletin of the Atomic Scientists* (November 1985): 24, 28.

103. "Excerpts from Speech By Gorbachev About Iceland Meeting," *NYT*, 15 Oct. 1986, p. A12.

104. Walter Pincus, "Soviet Says Talks Needed," *Washington Post*, 30 Oct. 1986, p. 4.

105. Philip Taubman, "Shevardnadze Specifies Limit on Star Wars Test," *NYT*, 11 Nov. 1986, p. A8. Moscow also released the text of proposals tabled at the follow-up Shultz-Shevardnadze meeting in Vienna. With regard to the Treaty, the Soviet proposal followed the Reykjavik language closely, with added clarification:

> Concerning the treaty on the limitation of ABM systems: Agreement to the effect that in order to strengthen the regime of the 1972 treaty on the limitation of ABM systems, which is of unlimited duration, the USSR and the USA will for 10 years not exercise their right to withdraw from that treaty and within that period will strictly observe all its provisions.
>
> Testing in space of all space-based elements of anti-missile defense will be prohibited, except research and testing in laboratories [*sic*]. This would not entail a ban on the testing of fixed land-based systems permitted under the ABM Treaty, or their components.
>
> Subsequently, within several years the parties will have to negotiate further mutually acceptable solutions in this area.

NYT, 7 Nov. 1986, p. A11.

106. R. Jeffrey Smith, "USSR Said to Request Parley on ABM Limits," *Boston Globe*, 26 Apr. 1987, p. 1. Monitoring satellite launches or agreeing on their contents raises a host of other problems regarding classified military payloads that are not ABM-related.

107. Stephen M. Meyer, "The U.S. SDI and Soviet Defense Policy: Near-Term Impact and Responses" (Paper delivered at the Aspen Arms Control Workshop, Aspen, Colo., July 1986). Cited with permission.

108. U.S., Congress, Senate, Armed Services Committee, *Hearings on the Military Implications of the SALT Agreements*, 92d Cong., 2d sess., 18 July 1972, p. 377.

109. U.S., Department of State, *Permitted and Prohibited Activities Under the ABM Treaty*, Current Policy Rept. 886 (Washington, D.C.: Department of State, Bureau of Public Affairs, November 1986).

110. Rhinelander, "The SALT Agreements," note 40, p. 128; Garthoff, "Correspondence," note 76, p. 108.

111. *1986 SDIO Report*, note 38, pp. C12–13 and VII-B-3.

112. Ibid., p. C7.

113. *Boston Globe*, 14 Sept. 1985, p. 6.

114. Ashton Carter, "The Relationship of ASAT and BMD Systems," *Daedalus, Weapons in Space, Vol. I: Concepts and Technologies* (Spring 1985): 171-89; William Durch, *Antisatellite Weapons, Arms Control Options, and the Military Use of Space* (Report prepared for the U.S. Arms Control and Disarmament Agency, Contract No. AC3PC103, November 1984), pp. 15-18.

115. *1986 SDIO Report*, note 38, p. C13; John H. Cushman, Jr., "Test on Missile Defense is Hailed by the Military," *NYT*, 12 Sept. 1986, p. D19; R. Jeffrey Smith, "SDI and Compliance," *Washington Post*, 15 Sept. 1986, p. 1; and Craig Covault, "SDI Delta Space Experiment to Aid Kill-Vehicle Design," *Aviation Week and Space Technology* (15 September 1986): 18-19.

116. On early Congressional efforts to limit ASAT testing and on the Soviet test moratorium, see William Durch, "Verification of Limitations on Antisatellite Weapons," in William C. Potter, ed., *Verification and Arms Control* (Lexington, Mass.: Lexington Books, 1985), p. 83. For fiscal 1985, Congress banned ASAT tests unless the president certified that the United States was "endeavoring in good faith to negotiate" ASAT limits; that testing against objects in space "is necessary to avert clear and irrevocable harm to the national security;" and that such testing would be "fully consistent with the rights and obligations of the United States under the 1972 anti-ballistic missile treaty. . . ." (From Sec. 1235, Public Law 98-94.) On August 20, 1985, Reagan so certified, and on September 13, the U.S. Air Force destroyed a U.S. scientific satellite in low earth orbit. *NYT*, 21 Aug. 1985, p. 10; Fred Kaplan, "U.S. Space Weapon Destroys its Target," *Boston Globe*, 14 Sept. 1985, p. 1. Thereafter Congress enacted a moratorium on testing the Space Defense System against objects in space, without the certification loophole. For descriptions of the Soviet co-orbital ASAT system, see John Pike, "Anti-Satellite Weapons and Arms Control," *Arms Control Today* (December 1983): 4-5; also Nicholas Johnson, *The Soviet Year in Space* (Colorado Springs, Colo.: Teledyne-Browne Engineering, 1981 annually forward).

117. Wilkening, "Space-Based Weapons," note 13, p. 149.

118. Clarence Robinson, "BMD Homing Interceptor Destroys Reentry Vehicle," *Aviation Week and Space Technology* (18 June 1984): 19-20; also *SDIO Report*, note 38, pp. C14-C15.

3 FUTURE PATHS FOR POLICY

Starting from a position of near-zero effective BMD—the legacy of the ABM Treaty—policymakers face an unaccustomed breadth of policy options, having different implications for the Treaty and for strategic defense. The most important questions for policy are not whether BMD can be made to work in some narrow sense (i.e., whether a bullet, or a beam, can hit another bullet), even though workable technology would be necessary to implement some options. The key questions are, rather, whether BMD is *needed* for some purpose (to replace deterrence, to enhance deterrence, to protect against accidental war, or to guard offensive force reductions); whether the costs and consequences of various sorts of BMD are acceptable (in terms of military spending, crisis and arms race stability, alliance cohesion, and East-West relations); and whether, given Soviet policy and action, we have any choice in the matter.

These questions all can be recast in terms of U.S. policy toward the Treaty; that is, in terms of how and whether the United States should aim to alter the political/strategic status quo. There are three basic policy alternatives: transition to a new offense/defense regime, deploy force defenses or other BMD of limited aims, or continue the current regime of near-zero BMD. The requirements, prospects, and likely consequences of each alternative are examined below.

TRANSITION STRATEGIES

There are two basic strategies for attempting to alter the current balance between strategic offense and defense. The first is competitive, with termination of the Treaty accepted as part of the price for building "defense dominant" East-West relations. The second strategy is a cooperative, bilateral effort, phased and regulated by arms control, in which the Treaty plays a role.

Competitive Transition/Termination

Willingness to compete toward a defense-dominant strategic world is implicit in Reagan administration arms control policy. Otherwise, the Soviet Union could "veto" the President's SDI vision. This implies in turn a willingness to withdraw from the Treaty should agreement with Moscow on a timetable for SDI testing and deployment prove unachievable. There are, however, a range of complications in following a competitive transition strategy.

Near-Term Policy Choices. Withdrawal from the ABM Treaty would be a more complex and risky operation than cancelling informal observance of an unratified agreement, such as SALT II. The Treaty is both in force and enjoys more political support than does SALT II. It is seen to impose significant limits on Soviet arms, whereas SALT II is seen to codify current arsenals. As the only bilateral agreement currently in force that constrains superpower strategic arms, the Treaty is seen as the linchpin of détente (still alive in Europe) and as a foundation on which to build future arms control agreements. U.S. withdrawal absent a material breach of the Treaty by the Soviet Union would reflect badly on America's other treaty commitments, both trade and security.

Advocates of U.S. withdrawal might build a case based on Soviet LPAR construction, arguing that such radars amount to a base for nationwide defense. But the case would not be clear cut; there would be Congressional and NATO opposition; the policy's impact on the 1988 U.S. presidential campaign could be counterproductive; and less drastic policy alternatives might better serve Western interests. That being the case, the broad interpretation of the Treaty serves as a near-term surrogate for withdrawal, giving high-technology development programs room to grow while constraining Soviet programs and postponing any final decision on the Treaty.[1] Nonetheless, to minimize the risk that the next U.S. administration would alter its strategy, the Reagan administration needs a near-term deployment policy that the next administration would feel constrained to carry out. To implement such a policy requires Congressional consent; such consent is unlikely absent compelling evidence of a Soviet breach. Although the Soviet Union is more likely to react materially to a near-term U.S. deployment threat than to a long-term research program, it is unlikely to react other than verbally and diplomatically to a policy that Congress had not sanctioned. Moreover, the Soviet decisionmaking window for the 13th Five Year Plan (1991–1995) will still be open in November 1988, meaning that Soviet leaders may make no resource commitments until the tenor of the next U.S. administration can be gauged.[2]

In short, there may be no viable near-term U.S. policy options available to implement a competitive transition strategy in a politically cost-effective manner.

Longer-Term Objectives. The prospect of population defense is what gives SDI its high standing in public opinion. Some members of the Reagan administration have set lower goals for SDI, casting its purpose in terms of damage limitation or, lower still, of enhancing deterrence only. The lack of a need for strategic defense to enhance deterrence was discussed previously. The need, feasibility, and desirability of BMD to limit damage in nuclear war, up to and including "assured survival," is the subject of discussion here.

Competitive Survival. To protect people, cities, and industries from nuclear attack, a missile shield would need to exceed 99 percent effectiveness against the *current* Soviet arsenal of ballistic missile warheads and cruise missiles. No defense system in history has been capable of such performance. At peak efficiency in the Battle of Britain, the RAF Fighter Command destroyed only 8 percent of incoming Luftwaffe bombers. In the 1950s, simulated Strategic Air Command (SAC) "raids" on U.S. cities suggested that ground-based U.S. air defenses had an effectiveness only of about 6 to 8 percent.[3]

Reliability problems alone would hamper efforts to achieve a near-perfect defense. It is very difficult to keep 85 to 90 percent of a high-technology force mission-ready at all times. The upper limit of readiness is also the upper limit of potential mission effectiveness, unless there is a considerable surplus of defense units deployed that can be programmed to substitute for defective units (25 percent over-deployment, for example, would boost a nominal 75 percent readiness rate to about 93 percent). However, this strategy would be especially expensive in the case of space-based systems, where the "absentee ratio" dictated by the shape of the earth and orbital mechanics would mean deploying ten to twenty units (depending on their effective range) to have one additional unit within range of Soviet ICBM fields at all times.[4]

These factors have led defense planners to favor "layered" systems, which would give a defense system more than one chance to shoot down a given warhead. Each successive opportunity reduces that warhead's statistical chance of survival. Passing through three defense layers, each with a kill probability of 75 percent, would reduce a warhead's probability of survival to below 2 percent.[5]

In that calculation, the capabilities of the first layer, the "boost phase," are critical, as is the nominally independent operation of the successive layers. Boost-phase defenses would engage missiles before warhead release; every MIRVed missile thus destroyed would mean many fewer warheads (ten in the case of the Soviet SS-18) and warhead decoys for mid-course and terminal defenses to discriminate and engage. Boost-phase "leverage" is thus potentially very high. However, by the same token, if boost-phase defenses fail to perform as expected or if they prove to be more vulnerable to countermeasures than expected, the overall performance of the defense system rapidly decays, as the warheads and decoys missed by the boost phase overwhelm subsequent layers. The offense need not double or triple its number of boosters to do so. As Peter Zimmerman notes, the offense need only push a higher-than-expected fraction of its force through the boost phase. That fraction may consist of covertly deployed boosters, or boosters with defense-penetrating capabilities (e.g., "fast burn" first stages and hardened upper stages). If "leakage" through the boost phase were expected to be 10 percent, increasing it to 20 percent would double the number of targets in the midcourse layers; increasing it to 30 percent would triple midcourse targets. If midcourse defenses cannot handle triple the expected raid size, the overflow would pass into the next layers of defense, overwhelming them in turn. In Zimmerman's scenario, a defense designed to allow

2 warheads to reach their targets could allow more than 800 to penetrate, more than enough to meet the criteria for assured destruction.[6]

In this system the defense layers do not operate independently;
their respective leakage rates are conditional upon the leakage rates
of preceding layers. Each depends on the success of preceding layers.
Making each layer of the system independently effective would
require separate command and control (C^2) systems with differing
failure modes (e.g., susceptibility to different sorts of countermeasures) and with kill capability sufficient to guard against the catastrophic failure of preceding layer(s). Such capabilities might be technically feasible, but would also be very expensive. Since the attrac-
tion of a layered defense lies in its theoretical ability to exploit the
leverage of the boost phase against MIRVed missiles, additional
boost-phase interceptor satellites might be orbited to compensate for
a "greater-than-expected threat." But as in the case of reliability
spares, adding one unit of reserve requires ten or more units in orbit.
Moreover, against the countermeasures threat, the defense can never
be certain of functioning as planned, since it cannot be tested against
the full capabilities of the real Soviet offense.

To cope with the countermeasures threat, the defense could concentrate efforts on the terminal layer, taking advantage of atmospheric friction to filter out decoys. But if those boosters carrying
maximum countermeasures are also most likely to cascade through
the defense, then the offense has some opportunity for "preferential
offense." To maintain a near-perfect defense that protects both civilian and military targets from a preferential offense wherever it might
be concentrated would require many thousand terminal defense
interceptors.[7]

The possibility of catastrophic boost-phase failure, and the cost
and difficulty of insuring against it, imply that defenses alone cannot
be relied upon exclusively to take the United States fully out of
pawn to Soviet nuclear power. The same applies to the Soviet Union.

Competitive Damage Limitation. SDI proponents argue that a
less-than-near-perfect defense could still thin out and break up counterforce attacks that require precise timing or precise laydown patterns, and perhaps limit damage from countervalue attacks. That is,
if SDI could not replace deterrence, it still could serve to limit damage should deterrence fail. But such defenses are likely to reinforce
targeting of leadership, C^3, war production, and economic recovery

capabilities, because precision targeting of hardened forces would be difficult, except under the most narrowly focused of counterforce options, in which a large number of warheads could be concentrated on saturating defense of a relative handful of targets (one kind of preferential offense).[8] Ten percent of the current Soviet strategic arsenal encompasses upwards of 1,000 warheads, or 500 to 600 equivalent megatons (EMT) of destructive power, indisputably sufficient to cripple or destroy the United States as a functioning society.[9]

Either side still might aspire to build a real damage limiting defense capability. However, such an effort would pose a direct challenge to the other's prevailing military strategy. Western strategy is designed to pose the clear risk that Soviet aggression will result in retaliatory damage that greatly exceeds any benefit derived from aggression. Neither U.S. liberals nor conservatives are eager to see the Soviet Union freed from that risk, a risk assumed to be at the heart of deterrence. Soviet doctrine, in turn, requires that U.S. forces and war-supporting infrastructure be placed and held at risk, and Soviet military leaders thus far give every indication that they intend to uphold this doctrine.[10]

Here are the makings of a competitive transition conundrum. The Reagan administration argues, consistent with current strategy, that continued offensive modernization is needed to protect the U.S. deterrent from erosion until the transition to defense is complete.[11] Under a competitive scenario, the Soviet Union could be expected to take similar steps to prevent the erosion of its offensive forces. Dual efforts to protect offense-based deterrence (that is, to keep the other side at risk) would tend to keep a transition on dead center while both sides' offensive and defensive forces increased.

If the United States were first to achieve a reversal in offense/defense cost-effectiveness (which currently heavily favors the offense), the Soviet Union would find itself in a double bind. Both its offensive and defensive counter-deployments would be more costly at the margin than U.S. options, because U.S. defense would be cheaper than Soviet offense, *and* U.S. offense would be cheaper than Soviet defense until the Soviet Union made its own defense breakthroughs or received the technology from us. The latter prospect is extremely remote. In sharing its technology, the United States would be handing the Kremlin the keys to the countermeasures. Moreover, if the United States held onto its breakthrough technology, it would be on the road to (temporary) nuclear superiority. Some analysts argue that this is just what the West needs and should seek, as a means of compelling Soviet cooperation.[12]

Competitive Endurance. If it were within U.S. capacity, would it be in U.S. interest to so corner the Soviet Union? There is a school of thought that suggests the Soviet political and economic system is sufficiently shaky economically, given its current high burden of armaments, that it might not be able to withstand the stress of a high-technology arms race, and that it then would foreswear offensive weapons and embrace both arms control and strategic defense.[13] Those analysts and officials who have for some years pressed the case about the relentless nature of the Soviet military buildup contend that a strapped Soviet Union cannot or will not build what it takes to counter a vigorous U.S. deployment of BMD. Yet a hard-pressed Soviet system, armed to the teeth and facing what it perceives to be the ultimate challenge from encircling imperialism, is unlikely suddenly to turn defeatist. Rather, Soviet military and diplomatic history suggest that under external pressure the Soviet response would be to redouble its efforts, not to seek accommodation. Moreover, recent U.S. intelligence assessments conclude that in an economic crunch the Soviet military sector would be sustained at the expense of other sectors without fatal economic consequences.[14]

Soviet prospects for sustaining the necessary defense spending in the long term may be better than those of the United States. The Reagan buildup saw the longest sustained series of peacetime defense budget increases since the United States first began to support large standing forces. But after four years of rapid growth, Congress began to cut the defense budget in real terms. SDI budgets have continued to increase, but at decreasing rates. With Congress now under Democratic control, future funding increases for SDI are problematic.[15]

Funding considerations aside, a two-decade commitment to develop and deploy SDI would span five Congresses (ten elections in the House and three for each Senator), five presidential elections, and three to five administrations (at least one or two of which would, based on historical patterns, belong to the loyal opposition). It would be highly undesirable if the United States and the Soviet Union both were to embark upon deployment of strategic defenses and offensive countermeasures, and the United States ran out of money, lost its will to compete, or both somewhere along the way. Alternatively, the United States could find itself locked into a competition that it had come to view as a mistake, but one to which the Soviet Union had become wedded with all the bureaucratic momentum at its disposal.

This is not meant to suggest that the United States is incapable of sustaining a defense policy over the long term. It has maintained significant strategic offensive forces and a significant troop presence in Europe and Northeast Asia for more than thirty years. Those policies and deployments, however, have benefited from a consensus view that they are necessary to ensuring the peace and prosperity that the United States and its allies have long enjoyed. There is no such consensus, either within the United States or within NATO Europe, that territorial missile defense has an analogous contribution to make. Until or unless such consensus is reached, the competitive path to strategic defense will remain at best politically uncertain.

Stability and Vulnerability. Assume that U.S. technology can create, its economy can finance, and both domestic and alliance politics will sustain a defensive system that promises to limit societal damage, perhaps to very low levels, in the event of nuclear war. Would the Soviet Union permit unilateral U.S. deployment of such a system? If so, could the United States sustain its defense dominance?

It has been argued here that strategic ballistic missile defenses are unlikely to prove sufficiently effective to take the United States out from under the Soviet nuclear threat unilaterally. But if the Soviet Union lacked confidence in the ability of its offense to penetrate the containment dome being erected over it, and if the expected price for crippling the dome preventively were lower than the expected price of inaction (which in this scenario would be neutralization of the Soviet deterrent), the Soviet Union might take steps to prevent the system's completion. The risk that such conflict would cascade earthward could deter such attacks, but the greater the projected effectiveness of the U.S. system (i.e., the more it appears that a competitive transition might succeed for the United States alone), the greater the Soviets' perceived risk of inaction.

Assume that a preventive attack did not occur, that both sides deployed extensive missile defense systems, and that the Soviet Union deployed at least some comparable types of weapons in space (BMD and/or ASAT). Assume, even, that space-based defenses were deployed under a cooperatively phased schedule (since the following discussion could apply to either the competitive or cooperative scenario). Space is generally a poor wartime environment for the defense. Satellites follow predictable paths and are exposed and visible to radar and infrared detection and tracking, particularly in the low earth orbits that "battle stations" would occupy. Their paths would carry them over Soviet territory, where ground-based efforts to disrupt the defense would be easiest to undertake. And they would be targetable by other space-based weapons. An orbiting BMD weapon would make a very good ASAT. Orbital lasers capable of burning or shock-fracturing a hardened missile body should be capable of doing the same to another orbiting weapon, sensor, or mirror at a similar range if position data were available (and no space-based defense could afford to forego the best possible intelligence on the other side's whereabouts).[16] In space-to-space warfare, a laser battle station would be an ideal prompt counterforce weapon, useful for punching holes in the other side's boost-phase defenses without prior warning. Defense suppression using space-based rocket interceptors would take a good deal more time, and would give some tactical attack warning. Infrared guidance sensors suited to homing on booster flames would be unsuited to attacking much colder satellites, but the system's self-defense components or its mid-course interceptors (designed to shoot at equally cool warheads) might be used. In short, like fixed, land-based ICBMs, space-based systems could engage one another directly. Fast, powerful, and relatively vulnerable, they could engender fears of preemption, suggesting the value of automatic shootback, which could in turn increase the risk of accidental engagement, giving the world the very best and most expensive fireworks display ever witnessed. When finished, it would leave the superpowers with full offense intact, truncated defense, problematic space-based communications, several thousand tons of high-technology junk in orbit, and great anxiety about what might happen next.

In other words, it would be difficult to keep a competitive transition "transitioned." As Glenn Kent and Randall DeValk, and Wilkening and others note in their studies of several notional competitive transitions, defense suppression capabilities would occupy a crucial niche in the nuclear arsenals.[17] A defense/defense suppression competition would supplement the expected competition between defense and responsive offense. In such a dynamic situation, as Glaser notes, conservative force planning principles and the inability to engage in full system tests would tend to leave both sides feeling inadequately defended. Uncertainty about the capabilities of their defenses would lead to requests for additional defenses, but meeting those requests "would yield little satisfaction and add little to the public's sense of security."[18]

Cooperative Transition

If the competitive transition is infeasible, or if feasible would produce undesirable results, how does the cooperative route fare? Damage limitation might be sought through offensive force reductions alone ("virtual defense") or through defense-supported reductions ("virtual transition"). Each approach is examined below.

Virtual Defense. Cutting the other side's offensive capability by a given percentage would buy as much "missile defense" as would a BMD system of equivalent "defense potential" (or ability to absorb warheads), but without incurring the costs of defense hardware and its modernization. The equivalent of uncertainty about actual defense effectiveness would be uncertainty about the amount of illicit offense that the other side might be able to deploy without detection. The risk in both cases would be couched in terms of the number of penetrating warheads expected in the event of war.

As a substitute for the counterforce-breakup capability of a very effective area defense, a virtual defense scheme would need to attend to the survivability of the remaining offense. Survivability of land-based forces would be enhanced if the number of launchers as well as the average number of warheads per missile were reduced, and if higher payload (i.e., higher "throwweight") missiles were dismantled at least as fast as the rest of the force. The launcher/throwweight reductions would progressively reduce the feasibility of barrage attacks against mobile, land-based missiles, since barrage capability (the ability to cover an area with a given blast overpressure) scales roughly with EMT, and deliverable EMT scales roughly with throwweight.[19] It would also allow mobile missiles to operate more securely in smaller operating areas, with potential benefits for verification.

Reducing the average number of warheads per launcher ("deMIRVing") would be the virtual equivalent of boost-phase defense. Although not so critical as boost-phase defense to its strategy's success, deMIRVing would reduce the ratio of weapons to hardened targets, reduce the feasibility of preemptive strikes against such targets (forces and C^2), and thus contribute to crisis stability.

Under virtual defense (with strategic forces at, say, 10 percent of their original strength), U.S. strategic forces could still pose a deterrent threat to Soviet nuclear escalation in Europe and could continue to back NATO's strategy of flexible response by keeping the Soviet homeland at risk. Colin Gray argues that such threats today lack credibility due to U.S. vulnerability, and that in a competitive loss-taking contest, the Warsaw Pact nations would win hands down.[20] U.S. vulnerability would not change appreciably under the virtual defense scenario, so the same criticism applies, as it would to a situation in which the Soviet Union were no longer vulnerable to U.S. nuclear threats. The same solution (NATO conventional force improvements) may also apply, unless the spirit that produced strategic force reductions also extended to reductions of shorter range ballistic missiles and ground forces in Europe.[21] In any case, the total cost of virtual defense plus conventional upgrades would be much less than the cost of conventional upgrades plus extensive BMD against current levels of Soviet offense.

Only incremental shifts in nuclear doctrines and targeting strategies would be required initially to implement this reductions strategy (say, through the first 50 percent of reductions). However, steady reductions over a long period would require wholesale changes in U.S. and Soviet force acquisition and targeting strategies, away from their current emphasis on prompt, hard target counterforce. Alternatives would be slow, hard target counterforce (i.e., increased reliance on aircraft and cruise missiles); other counter-military targeting; and economic/industrial targeting (i.e., assured retaliation/assured destruction). Reducing all offensive forces still further would require even more wrenching shifts in doctrine and strategy, as nuclear deterrent threats dwindled in size and as concern focused on the potential utility of marginal, covert additions to residual forces.

The period of doctrinal adaptation and the record of cooperation that a reductions regime established might facilitate negotiation of the more radical cuts. It might also facilitate negotiation of missile defenses to "guard" the resulting regime.

Virtual Transition. Kent suggests that nationwide BMD might be deployed to guard against cheating when reducing strategic ballistic missile forces "to zero, or near-zero" levels. Kent's proposal, like all other proposed transitions that incorporate arms control, is a "virtual transition" scenario because its success depends on a political settlement and jointly agreed-upon behavior, not on the unilateral triumph of political will and defensive technologies. Defense deployment in Kent's proposal would be phased to begin after 50 percent cuts in offensive forces had been made, and would be sized to defend against cheating on reductions as ballistic missile forces approached total elimination.[22] Final offensive missile reductions would be made rapidly to minimize potential instabilities as offensive force size passed through the region where deployed BMD might support a preempt-and-defend (what Kent and DeValk term a "conditional survival") strategy. In Kent's formulation, deterrence would be reinforced throughout the final stage of ballistic missile reductions by the bomber/cruise missile force. He also stresses the importance of the proper phasing of reductions and defenses: "To preempt this agenda by prematurely deploying defenses in the absence of any agreement for cooperation would serve to preclude any hope of ever pursuing the only approach [to assured survival] that seems plausible. To change from cooperation to competition is always possible. The reverse—from competition to cooperation—may not be that easy."[23]

Kent argues that neither side should object to deployment of defenses against a class of offense that is about to be eliminated.[24] Yet BMD in a virtual transition need not, and should not, be capable of more than guarding against cheating on a zero-ballistic-missiles regime. BMD sufficient to defend against a large-scale attack might disrupt rather than facilitate second-phase offense reductions, and if built up after such reductions were complete, could give a side the foundation of a first-strike capability should it later choose to quickly rebuild its offense.

The two sides might defer detailed consideration of BMD deployment ceilings until after the first stage of offense reductions was well underway. A successful record of implementing reductions might facilitate subsequent negotiations on defenses.[25] Without full Soviet cooperation in amending the ABM Treaty, however, the virtual transition phased in this manner would either stall (leaving offensive forces at 50 percent of current levels), shift to virtual defense (decreasing ballistic missile forces without BMD), or default to a competitive transition effort.[26]

All the provisions of this decade or longer venture should therefore be set upfront, including any changes in the ABM Treaty. Should implementation of *this* arrangement be disrupted for some reason—doctrinal, perhaps, or political—the sides' commitment to move defense components rapidly through development and into production could carry one or both sides' BMD systems into deployment, producing what the virtual transition tries to avoid: significant offense and partial area defense.

Assuming that implementation went smoothly, the anti-cheating BMD system would need interceptors with exo-atmospheric range and sophisticated on-board tracking and guidance capability or support from an extensive battle management radar network. It might also be designed as a "cadre" system with a rapid deployment capability to counter offense cheating by the other side. In either case, the system would seem inherently expandable and thus could contribute to the breakout problem it is supposed to resolve, unless cheating on defense limits were as readily detectable as cheating on offense limits.

A side's excess production of BMD components might not be a cause for concern if offense limits could be reliably monitored (since the offense would be necessary to make this sort of breakout scheme threatening to the other side). But the anti-cheating BMD system is premised on *un*reliability of such monitoring of the offense. If such monitoring were reliable, presumably there would be no need for a BMD-based insurance policy because offense cheating would be detected in time to be offset other ways. Therefore, potential cheating on BMD limits would be a cause for concern, and monitoring capabilities would need to be good enough to keep close track of defense deployments and production.

Yet if capabilities were good enough to monitor compliance with partial constraints on BMD, they ought to be good enough to monitor compliance with a ban on strategic ballistic missiles. Keeping track of potentially mobile BMD components and assuring their limitation when production lines are warm is almost by definition harder than monitoring a ballistic missile ban, where any detection constitutes violation and production facilities may be shut down and monitored for activity. Thus, if monitoring were good enough to verify limitations on its deployment, BMD in a virtual transition would seem to be unnecessary.

The risk of breakout would be greatest in a world in which air-breathing offense and corresponding air defenses had also been dismantled (Schelling's classic unstable disarmed world).[27] Ballistic-missile breakout concerns could be alleviated, on the other hand, if bombers and cruise missiles were not included in the final reductions in offensive forces. The sides might have little choice in the matter, since eliminating air-breathing strategic forces in the phased and defense-protected manner envisioned for ballistic missiles may not be feasible.

Because the Soviet Union already possesses a significant air defense, turning that system into an arms control insurance policy would require considerable reductions in its capabilities. But much of the Soviet air defense system is multi-functional, serving tactical as well as strategic purposes and guarding against conventional as well as nuclear attack, so Moscow is highly unlikely to scrap most of it in the context of a strategic arms agreement. Moreover, the difficulty of verifying the elimination of strategic cruise missiles (because they are small, and because the same type of airframe can be fitted with a variety of warheads and used for a multitude of non-nuclear missions) is likely to reinforce that reluctance, since an adequate level of defense insurance would be hard to calculate. It would be equally difficult for the United States. The strategic value of a covert force of cruise missiles could be quite significant under circumstances in which the cruise was both sides' main strategic weapon. Thus, a bomber/cruise missile elimination agreement could be quite risky. On the other hand, cruise missiles would be relatively cheap to produce quickly in response to a breakout effort and, of all air-breathing forces, they would be most likely to penetrate an air defense, which would reduce the risk of a reductions agreement but not improve its stability or verifiability. Trying to reach mutually assured survival by closing off air-breathing offense as well as ballistic missiles is thus unlikely to succeed.

If the threat from a major class of nuclear weapons (aircraft and cruise missiles) cannot be eliminated reliably and verifiably, then eliminating strategic ballistic missiles becomes, as Kent notes, "not an intermediate step along the way" to a larger goal but "the end in itself." This in turn "casts a pall over" proposals to eliminate ballistic missiles, "at least from the standpoint of [achieving] assured survival."[28] This then casts doubt on the desirability of undoing the ABM Treaty on behalf of a virtual transition scenario. The continued existence of an air-breathing offensive deterrent would reduce breakout incentives in a virtual transition, and mitigate the impact of a breakout attempt, but that would seem to be the case whether or not BMD "guards" the missile reductions regime.

Summary

This review of the requirements and implications of transition strategies suggests that a number of unpleasant consequences would flow from pursuing a competitive strategy. Damage limitation, let alone assured survival, would be extremely difficult to achieve under conditions of open competition between offense and defense, defense and defense suppression. If achieved, either capability would be difficult to sustain; indeed, achievement of either one could never be demonstrated in the absence of war. The uncertainties attending the capabilities of the defense might make a would-be aggressor hesitate to attack, but the source of that hesitation—fear of devastating consequences—would be no different than today, despite a very large investment of money, material resources, and national psyche.

Under the cooperative strategy, offensive reductions would buy a degree of "virtual defense," that is, a threat reduction equivalent to that provided by a given level of BMD. They would not remove the Soviet nuclear threat entirely, but neither would an equivalent BMD system. BMD might insure the stability of a "virtual transition" agreement that reduced strategic ballistic missiles to zero, but only if the defense could be verifiably constrained. Operating on a nationwide basis, it would be more difficult to monitor than are current Treaty limits, and its permitted deployments would be more expandable. Since air-breathing offense appears difficult to reduce reliably to near zero, the net result would be, not assured survival, but a modified deterrent regime that required the United States to give up, among other things, its strategic missile submarine force, arguably the most secure and least crisis-destabilizing element of the deterrent.

LIMITED DEFENSE

The second of the three policy paths involves modifying the ABM Treaty to allow deployments of BMD for limited aims, such as defense of strategic forces, particularly ICBMs, or defense against accidental or "Nth country" attacks.[29] The following discussion assumes levels of offense that may reasonably be expected to prevail in the mid-1990s, extrapolating from current and planned forces absent reductions due to arms control, except where noted, and absent crash building programs. (The mid-1990s are the earliest time by which non-nuclear terminal defenses may be available for deployment.)

Limited BMD is first discussed in terms of U.S. and Soviet strategic interests; second, in terms of requirements for defense of ICBMs (before and after offense cuts); third, in terms of defense against accidental or Nth country attack; and fourth, in terms of required changes to the Treaty.

U.S. Strategic Interests and Limited BMD

Current U.S. nuclear strategy seeks to maintain a survivable retaliatory capability sufficient to cause unacceptable damage to the Soviet Union over a wide range of scenarios, "prevailing" in a protracted conflict if necessary.[30] It is generally believed that BMD could facilitate this strategy and bolster deterrence by improving the ICBMs' prelaunch survivability.

Those who conclude that hardsite defenses are necessary tend to view land-based ICBMs as a crucial element of U.S. nuclear capability and their vulnerability as a crucial chink in America's deterrent armor. Those who do not consider hardsite necessary view ICBM vulnerability as an overrated problem, fixed-silo ICBMs as expendable or replaceable, and limited, bilateral BMD as a net gain for the Soviet Union. These positions are considered in turn.

In its report to the president, the Scowcroft Commission notes that, in comparison to other legs of the U.S. strategic Triad, ICBMs "have advantages in command and control, in the ability to be retargeted readily, and in accuracy." They "are especially effective in deterring Soviet threats of massive conventional or limited nuclear attacks, because they could most credibly respond promptly and controllably against specific military targets and thereby promptly disrupt an attack on us or our allies." The Commission also notes that the differing capabilities and vulnerabilities of the bomber and ICBM forces make it difficult for the Soviet Union to attack both successfully.[31] That is, if Soviet sea-launched ballistic missiles (SLBMs) and ICBMs were launched simultaneously, SLBM detonations on U.S. bomber bases and other targets would give the United States 10 to 15 minutes confirmation of attack before Soviet RVs reached American missile silos, such that Soviet planners could not rule out U.S. resort to launch under attack for the ICBM force. On the other hand, an attack using simultaneous weapon arrivals would launch SLBMs 10 to 15 minutes after ICBM launch and give U.S. bombers nearly 30 minutes' warning time in which to escape their bases. (In neither case, of course, would U.S. missile submarines on patrol be vulnerable to such attacks.) If Soviet SLBMs become accurate enough to attack targets as hard as missile silos, the Soviets' attack timing dilemmas would be reduced and all fixed, land-based U.S. targets would face a similar "failure mode." This may be an argument in favor of BMD, although it is equally an argument about the waning value of silo-based ICBMs and the growing value of mobility and concealment as defensive measures.

Even though the United States cannot be disarmed by Soviet attack, some analysts are concerned about a hypothetical Soviet attack against U.S. ICBMs that deprived the United States of prompt, hard-target counterforce capability and restricted U.S. retaliatory options to countervalue SLBM strikes and slow bomber counterforce strikes. Because the Soviet Union would have counterforce superiority and still hold U.S. cities hostage, it is argued, U.S. retaliation could be completely deterred.[32] Vulnerable U.S. ICBMs therefore invite Soviet efforts to achieve coercive advantages in crisis.

Albert Carnesale and Charles Glaser argue that a large-scale Soviet attack against U.S. ICBMs would not give Moscow a coercive advantage. Such an advantage could derive only from drastically reducing the absolute amount of damage that U.S. forces could inflict in retaliation. Since extensive U.S. forces would survive a Soviet counterforce attack, the absolute level of destruction deliverable to Moscow's doorstep would remain very high. Thus, striking the U.S. ICBM force would not be an attractive option for the Soviet Union in a crisis.[33]

Carnesale and Glaser note that the ability to hit *time-urgent*, hardened targets in a second strike is often considered to be a uniquely important capability of U.S. ICBMs, and the Scowcroft report implies as much. However, they find it difficult to identify time-urgent second-strike targets. Soviet reserve ICBMs might be such targets, but these could be launched upon warning of U.S. attack, obviously easier to do when one's forces are expecting attack than when they are attacked "out of the blue."[34]

Reloadable silos for "cold launched" Soviet SS-17s and SS-18s might also be considered priority ICBM targets. But reloading is estimated to take several days, during which time reload crews and equipment would be vulnerable to any nuclear detonation within a several-kilometer radius. Ten hours after launch, those silos could come under attack from accurate U.S. cruise missiles. SLBMs could also be used against road and railway routes to each of the eight SS-17/18 deployment areas, to disrupt reloading while the cruise missiles were enroute. In the 1990s, accurate Trident II SLBMs could be targeted against the silos themselves if the United States were seriously concerned about the reload problem, which, it should be stressed, does not contribute to Soviet first-strike capabilities.[35]

Other hard targets would be no more time urgent. Since the main Soviet attack would have already hit the United States, and since Soviet reserves could be primed to launch upon tactical warning of U.S. retaliation, hitting Soviet C^3 would do little to limit damage to the United States. The vulnerability of Soviet leaders in hardened shelters is unlikely to be greater thirty minutes into the war than eight hours later. In any case, lobotomizing Soviet C^3 capacities may not be the best route to early war termination, which is a stated U.S. strategic objective.[36] Soviet strategic forces could revert to any number of predetermined contingency plans if C^3 capacities were lost; wholesale standdown of reserve forces is only one possibility, and not necessarily the most likely.[37]

In short, although U.S. ICBMs are theoretically vulnerable to Soviet attack, such an attack is very unlikely. Moreover, should war occur, it is not clear what set of hardened targets would be so time urgent in a second strike as to require destruction in minutes. If prompt, hard-target retaliation were necessary, then by the mid-1990s accurate U.S. SLBMs should be able to stand in for attacked ICBMs.[38] Defense of ICBMs does not appear to be necessary for the implementation of U.S. nuclear strategy.

Glaser observes that if there were important, time-urgent, second-strike targets in the Soviet Union, deployment of Soviet BMD would make them more difficult to destroy.[39] There is the further difficulty that equivalent Soviet BMD capabilities, which one would expect to result from a mutually agreed deployment, would result in significant Soviet advantages relative to the current situation. These implications are discussed shortly.

Soviet Strategic Interests and Limited BMD

From the time of the SALT I negotiations to the present, the Soviet Union has evidenced little interest in active defense of its ICBM fields. Soviet strategy plans for preemptive use of its ICBM force or the launch of some fraction of it upon actual evidence of attack.[40] Yet in the 1970s the Soviet military invested in ICBM silo hardening to maintain their survivability as U.S. missile accuracies improved, and in the 1980s is turning to mobile ICBMs, as missiles such as MX and Trident II give the United States the capability to destroy the hardest of Soviet fixed targets with little warning. Soviet strategy does require secure reserve forces for follow-on strikes, and to hedge against the possibility that wartime complications, including failure to correctly anticipate an enemy attack, might inhibit implementation of all but a second-strike rideout, retaliatory policy.[41] Taken together, Soviet operational strategy and passive survivability measures may obviate the need for active defense of ICBM fields and permit a given investment in BMD to be directed to other ends.

Sayre Stevens suggests that, given a choice, the Soviet Union would assign highest priority to upgrading the defense of Moscow, followed by defense of "military and industrial concentrations in the western USSR." Defenses would aim to "preserve those elements important to warfighting and reconstitution" as part of an overall damage-limitation strategy.[42] Other objectives might include defense of the C^3 network. Because many of its command centers are blast-hardened and redundant, with much-less-than-perfect BMD they might be defended preferentially, much like ICBM silos.

Soviet BMD that had at least some area defense capability would better fit the array of missile threats that the Soviet Union will continue to face through the 1990s. The Chinese have a handful of ICBMs, which are unlikely to pose a threat to Soviet forces but which are capable of reaching Moscow. Barring fundamental changes in British defense policy, its Polaris system will be replaced in the mid-1990s by new submarines carrying MIRVed Trident II missiles. The changeover will increase on-station, independently targetable British warheads from 32 to 128 by the turn of the century. This will still be too few to support a hard-target counterforce targeting strategy but enough to do serious damage to the Soviet military and civil-

ian infrastructures, and perhaps to elements of command and control
as well. Until Trident is operational, British Polaris submarines will
carry the *Chevaline* multiple-warhead ABM penetration system.
France is following a similar strategy with deployment of the multiple-warhead M-4 missile aboard its strategic submarines; it will be
succeeded by a true MIRV in the mid-1990s.[43]

Soviet BMD might also contribute to the survivability of Soviet
air defenses, key parts of which would otherwise be destroyed quickly by U.S. missile strikes.

Thus, given the opportunity to deploy limited, negotiated amounts
of BMD, Soviet policymakers would be more likely to choose deployments able to defend C^3 and other military targets in a combination of point and area defenses than they would be to choose to
defend ICBM fields.

Defending ICBMs

This section examines three applications of BMD: to defend silo-
based ICBMs against a full-sized Soviet offense, to defend mobile
ICBMs against a similar threat, and to defend residual silo-based
ICBMs after deep reductions in both sides' forces have been taken.

Silo-Based, No Reductions. If the United States were seriously
thinking about deploying BMD for its strategic forces, it would not
want to proceed with negotiations to modify the Treaty (and permit
deployment of Soviet BMD) before knowing whether such deployments would really provide a net gain in U.S. retaliatory capabilities.
If a defended U.S. force, facing Soviet defenses, did not pose at least
as great a deterrent threat to Soviet targets as an undefended force
with a "free ride" to its targets, proceeding with BMD deployment
would be counterproductive.

Glenn Kent and Richard DeValk offer a set of simple equations
for calculating the approximate levels of BMD that are required to
produce a desired level of silo-based ICBM survivability, taking into
account the defense's strategy and the size of the attack. For sim-
plicity, the equations assume that each incoming RV has a probability of kill (PK) of 100 percent against an ICBM silo, and that each
BMD interceptor has a PK of 100 percent against an ICBM silo, and
that each BMD interceptor has a PK of 100 percent against an incoming RV. The defense is assumed to be survivable and to consist

only of terminal components (e.g., ground-based, endo-atmospheric interceptors and radars) operating in the "discriminating semipreferential" mode determined to be most effective for this scenario.[44] To enforce a 50 percent survival rate for 1,000 ICBM silos against a 5,000 RV attack would, under these assumptions, require 4,500 interceptors. If assumptions about the perfect operation of the defense were relaxed to make the engagement more realistic, roughly 8,000 interceptors would be needed to ensure 50 percent silo survival.[45]

Such a BMD system would require changes in Article III of the ABM Treaty to permit increased numbers of permitted deployment areas and greatly increased numbers of launchers, interceptors, and radars. Depending on the range capabilities of permitted sensors and interceptors, one ABM deployment area with a 150 kilometer radius (the size of currently permitted deployment areas) could be required for each of six wings of Minuteman/MX. (Four SAC bomber bases are co-located with Minuteman wings and would also fall within the system's "footprint.") The system would require a number of survivable radars for each deployment area. Radars and perhaps launchers would probably need mobility and/or deceptive basing for their own survival; this would require relaxation of the land-mobile basing mode restrictions in Article V. The potential utility of air- or space-based sensors in discriminating RVs from decoys and in cueing the defense would suggest relaxation of Article V's limits on ABM radar substitutes that are not fixed land-based. With the distribution of ABM radars around the country, limits on placement of missile early-warning radars (such as Krasnoryarsk) may become moot, unless ceilings on the power-aperture product (mean emitted power times transmitter area) of the newly permitted ABM radars were substantially lower than would be useful for area defense.

Finally, Article I's prohibition of a "base" for defense of national territory would need to be revised, unless permitted interceptor range capabilities were severely constrained. Especially if component mobility and some component space basing were allowed, constraints on kill mechanism range could be important in keeping defenses sufficiently constrained that value targets would remain difficult to protect. Such constraints might be based on the range capabilities of the U.S. Sprint or the Soviet SH-08 interceptors (roughly 50 to 75 kilometers). Without further constraints on mobility, assessing the number of deployed components could be extremely difficult, but countable fixed-based point defense may not be survivable.[46]

The deployment of so many interceptors and radars would likely fatally undermine what remained of the ABM Treaty regime. The United States is suspicious now of Soviet capabilities and intentions with respect to breakout. It would have far stronger grounds for concern about compliance, whether or not it was Moscow's intent to breach the new accord, if the sides were permitted to deploy nearly as many ABM interceptors as there are Soviet SAMs, and C^2 links between the BMD and SAM systems could prove impossible to prevent. The further expansion of such a system into a more complete area defense could be relatively straightforward: deployments might be made denser or more extensive on relatively short notice, and deployment of longer-range interceptors would not be difficult if production lines for Galosh or its follow-on remained open.

The question of Treaty impact is secondary, however, to the impact that deployment of up to 8,000 Soviet BMD interceptors could have on U.S. retaliatory capabilities. That number of Soviet interceptors could absorb nearly 80 percent of 2,900 hard-target-capable U.S. ballistic warheads (all but about 450) fired in retaliation, both ICBMs and SLBMs, if each Soviet missile defense interceptor has no more than a 50 percent chance of destroying a U.S. missile warhead. Soviet interceptor PKs would have to sink below 10 percent before the number of penetrating U.S. ballistic missile warheads would begin to approach levels expected in the current, undefended case.[47]

Why is this so? Silo-based ICBMs are the only segment of U.S. strategic forces whose prelaunch survivability is substantially improved by BMD. By 1996, the ICBM force will carry less than half of the hard-target-capable ballistic missile warheads. The other half will be aboard Trident submarines, whose missiles are not vulnerable to preemption when their ships are at sea, and which do not benefit from U.S. BMD. Nonetheless, their penetrability *is* eroded by Soviet BMD. The U.S. bomber force, whose substantial retaliatory capability is not considered in this analysis, could be affected at the margin by Soviet BMD as well. U.S. bomber bases, being large, soft targets, would be difficult to defend against Soviet attack, but Soviet BMD could blunt defense suppression attacks by U.S. ballistic missiles against Soviet air defense installations, leaving the U.S. bomber force to fight its own way through Soviet air defenses.[48]

Mobiles. A relatively modest BMD system might be deployed, alternatively, to defend the postulated 500 mobile Midgetman ICBMs.

The development and deployment of the system would require different changes in the ABM Treaty. One analysis suggests that Treaty-permitted *numbers* of ABM launchers and radars might be able to double the attack price against Midgetman, if the ABM components were land-mobile and if each defended a cluster of five Midgetman launchers against a uniformly distributed attack on the Midgetman deployment area(s).[49]

Such a system would require alteration of Article III of the ABM Treaty to permit new ABM deployment areas (as many as there are mobile ICBM areas or bases); and of Article V, to permit development, testing, and deployment of land-mobile components. The altered Treaty could retain the prohibition on a base for national defense if the defense were confined to short-range interceptors and radars appropriate to deep terminal defense, or if mobile ICBM deployments were confined by agreement to areas remote from major urban concentrations. Other Article V limits on ABM component mobility probably could be retained.

A well-dispersed force of off-road mobile ICBMs would be a difficult target set to attack; BMD might make the task more difficult. BMD also might operate with a good deal of leverage if location uncertainty of all mobile components (offense and defense) could be preserved.

If location uncertainty were preserved routinely, however, compliance with agreed BMD deployment limits would be harder to verify and might have to be supplemented by component production monitoring. Indirect counting measures, like the deployment "gates" planned for the multiple protective shelters MX basing scheme, imply physically bounded deployment areas that would be inconsistent with mobile ICBM survivability unless the areas were extremely large or the barrage threat against Midgetman were declining. (Of course, if the threat were declining, active defense of mobiles would be a wasting asset.)

Direct monitoring of BMD component production would be more useful, and probably advisable, to foreclose covert stockpiling of mobile components, which could support an expanded system suited to rapidly tapping whatever latent BMD capabilities resided in existing SAM and ATBM systems. Direct monitoring, coupled with deployment of mobile ICBM and BMD components in areas remote from major cities and industry, would minimize that risk. Such components might only be deployed, for example, within the rather large areas described in the following section, complementing offense and defense based deceptively in silos.

Silo-Based, After Reductions. Limited BMD might be deployed to enhance the survivability of remaining ICBMs after deep reductions in offensive forces. As reductions in offensive weapons proceed, some ICBM silos might be converted to accommodate ABM launchers and interceptors or their associated "pop up" radars, in a variation of the U.S. Sentry concept developed for defense of MX in multiple shelters. Emptied ICBM silos could otherwise be dismantled and destroyed according to agreed procedures, or be converted to hold canisterized ICBMs in a further variation on multiple shelter deployment. The defense might be similarly canisterized and moved periodically. To verify that agreed levels of defense and offense were not exceeded, all silos might be opened periodically for overhead inspection.

Since the purpose of missile defense in this case would be stability enhancement and not defense of national territory, ideally BMD for residual silo-based missile forces would be located away from urban/industrial concentrations and would use interceptors with limited range. That would not be too difficult in the U.S. case, if the core of land-based forces to be defended included the Minuteman fields at Grand Forks, Minot, Ellsworth, and Warren Air Force Bases, which currently incorporate 500 silo launchers and 3 major SAC bomber bases. Together, they occupy a triangular area of roughly 200,000 square kilometers largely within the borders of North and South Dakota and Nebraska, relatively far from the major urban concentrations of the Great Lakes area.

On the Soviet side, the equivalently remote ICBM deployment areas might include those east of the Ural Mountains at Uzhur, Aleysk, and Zhangiz Tobe, where approximately 200 SS-18s are currently deployed, and Gladkaya, an SS-11 main operating base with 75 to 100 launchers.[50] They occupy a triangular area of roughly 375,000 square kilometers. The nearest other ICBM deployment areas are 1,100 kilometers to the west and east. If the reductions scheme called for cuts in average throwweight and number of warheads per ICBM launcher, the SS-18 silos could be emptied and converted first.

This scenario would require changes in the number and definition of permitted ABM deployment areas. The size of the permitted ICBM defense site would have to be increased by three to five times and the number of permitted launchers would have to be increased as well. The absolute increase would depend on the number of ballistic missile warheads remaining after cuts and on the leverage afforded by deceptive basing. Since this BMD system would employ fixed-based components, it should not require changes to Article V or to other aspects of the Treaty. (If interlaced with mobile ICBMs and their own BMD escorts, then the above-mentioned changes to Article V would of course be necessary.)

It is easier to conceptualize a BMD system to cap a reductions regime than it is to devise a military requirement for it or to envision successful negotiations to permit its deployment. If the reductions agreement itself is a sound one, then it should have a reduced average-missile throwweight, a reduced average number of warheads aboard remaining ballistic missiles, and a reduced number of missiles. It should have deemphasized silo basing in favor of mobile land basing or basing on submarines. If it makes these reductions and permits multiple-shelter basing of residual silo-based ICBMs, then active defense of those residual forces should not be necessary to ensure adequate survivability. If not, then deployment of the very limited sort of BMD discussed here would not correct the damage.

With respect to negotiating such defenses, the Soviet Union has shown little interest in active defense of its ICBMs. As the Soviets move to mobile ICBM basing and deploy more survivable submarines and bombers with cruise missiles, that interest is unlikely to increase.

A further application of limited BMD involves the Sentinel rationale: missile defense against accidental launch and Nth country attack.

Accidental/Nth Country BMD

Once again, first look at the need for such a system, then at what it might require in terms of hardware and changes to the Treaty.

Paul Bracken discusses four separate scenarios for accidental/inadvertent nuclear war. The first is "pure accident," in which a missile is launched by human error or mechanical failure. The second through fourth scenarios involve a war arising from missteps made during conditions of prior nuclear alert, international crisis, or conventional war.[51]

Of these four scenarios, only the first is likely to produce an attack involving only one or several nuclear weapons. An accident-absorbing BMD deployment would be designed for that scenario. Bracken's other three scenarios involve nuclear *wars* that start because, after the system has been put in gear, someone's foot slips off the clutch. A BMD system to guard against damage in those three scenarios must be capable of limiting societal damage in a full-scale nuclear war. Discussion of that sort of system is found earlier.

However, the first scenario is also the only one Bracken considers to be a "virtual impossibility."[52] Since the early 1960s, the United States has implemented redundant warning systems and decision procedures; positive control procedures for missile and aircraft launch (no missile launch without explicit authorization, including launch codes for land-based forces and two or more physically separated humans in the final launch decision loop; no airborne alerts; and recallable launch of strip-alert bombers); weapons designs that self-disable if the weapon is tampered with; and "permissive action links" (or PALs, encoded electronic locks that separate "control of the weapon from possession of the weapon").[53]

The Soviet Union uses a combination of redundant warning, PAL-type electronic interlocks (apparently even on its missile submarines, which the United States does not) and political controls, including KGB monitoring of political and personal reliability, a high percentage of Communist Party membership among military officers with nuclear weapons-related responsibilities (80 to 90 percent versus 7 percent in the populace at large); and involvement of unit Political Officers in nuclear launch decisions.[54]

None of these features render the purely accidental launch of a strategic nuclear weapon totally impossible, but in combination they render its probability very low. Supporting respective unilateral procedures are the Moscow-Washington Direct Communications Link (the "Hotline") and the 1971 U.S.-Soviet "Accidents Measures" agreement, which commits the two countries to notifying each other in the event of "unexplained nuclear incidents," especially those that carry any threat of nuclear war (e.g., nuclear weapons detonation or detection of unidentified objects by early-warning sensors). Each party is also required to notify the other of planned missile flights that extend beyond its national territory "in the direction of the other party."[55]

The issue of Nth country attacks has been a rationale for BMD since the mid-1960s, but Nth country ICBM/SLBM capabilities have been slow to proliferate. The United States obviously would not consider building a BMD system to guard against French or British attack, and the only other country having long-range nuclear missiles is China. U.S.-China relations have come a long way since 1967, as Chinese domestic politics have stabilized and China has become a quasi-ally of the United States. In 1987, no BMD system could be sold to Congress on the basis of the Chinese threat, even though China now has a half-dozen ICBMs and some missile submarines.[56] There currently are no other non-Soviet long-range nuclear ballistic missile threats to the United States. To the extent that there are other Nth country threats out there, they are more likely to reach U.S. territory aboard ostensibly commercial ships and aircraft.

The Soviet Union is a more likely target of Nth country nuclear action, and partly for this reason maintains its Treaty-permitted Moscow ABM system. That system could also protect a substantial fraction of the western Soviet Union against accidental missile launches, although use of its nuclear-armed interceptors would carry serious consequences for the "protected" area.

If the United States wished to deploy its own protection against accidental attack, it might opt to deploy very long-range interceptors (such as ERIS) at its Grand Forks, North Dakota ABM deployment area. If this did not provide satisfactory coverage, the United States might seek to divide currently permitted components between two deployment areas, east and west, each having fifty interceptors. However, the Treaty would have to be amended to permit this option.

Summary

The case for defending U.S. ICBM silos against Soviet missile attack, cast in terms of the need for ICBMs and their current vulnerability, is not clear-cut even before assessing the net strategic impact of such defenses. However, it appears that deploying hardsite defense sufficient to protect a significant fraction of silos (up to one-half) would permit Soviet BMD deployments sufficient to reduce U.S. prompt hard-target retaliatory capabilities to about 20 percent of what they would be without BMD. Defenses of that magnitude could not be considered limited in Treaty terms, and a Treaty changed to permit them would be fundamentally different in purpose.

A system to defend only mobile ICBMs might do much less damage to the ABM Treaty, but it is not clear that mobile ICBM survivability would need such an assist, particularly in the context of offense reductions. Defense of residual silo-based ICBMs also need not pose a threat to the Treaty's viability, but would not seem necessary and probably would not be of great interest to the Soviet side.

The case for a BMD system to guard against Nth country attack is also found to be weak, but a system intended to guard against accidental missile launches might be deployed within the current terms and intent of the Treaty. Its cost-effectiveness is not discussed, but given the low probability of an accidental attack, it might be better justified from the U.S. viewpoint as a restorer of political symmetry (i.e., as a match for the Moscow system).

NEAR-ZERO BMD: STRENGTHENING THE ABM TREATY

The third path for future policy would continue the current regime of near-zero BMD and perhaps strengthen it to keep the ABM Treaty regime viable into the 1990s and beyond. The deployment and development issues discussed earlier would need near-term attention, as would public rhetoric about the Treaty.

Rhetoric and Policy

To preserve and perhaps strengthen the Treaty, it is not necessary that the two sides have identical purposes in mind. However, it is necessary that both wish to maintain an effective agreement that applies to them in equal fashion. The first indicator of serious interest in a strengthened Treaty would be a lowering of rhetoric about the anticipated contribution of BMD to nuclear strategy, and moderation in public assessments of research results. Recent Soviet statements have been supportive of the Treaty and of strengthening it, although Soviet behavior has sent mixed signals that have yet to be tested at the bargaining table. Increasingly, U.S. statements have supported the Treaty only in terms of the broad interpretation, while statements of support for early deployment of SDI have begun to supersede statements that SDI is only a research program. This U.S. posture must change before substantive negotiations to strengthen the Treaty can be undertaken.

A potentially important change to the Treaty may be an altered withdrawal clause. The short-notice clauses in this and other arms control agreements have functioned as domestic political buffers emphasizing that the United States can leave the agreements if necessary, thus lessening the sense that the United States has handed its security to its treaty partners. Of course, Western security rested in Soviet hands, and vice versa, long before SALT, and the current U.S. debate over SDI is really an argument over whether this is irretrievably true. The content of the withdrawal clause in the ABM Treaty has little bearing on that debate. But agreement to a longer-term withdrawal clause would indicate long-term political commitment to the Treaty.

In 1985, John Rhinelander, former legal advisor to the U.S. SALT I delegation, proposed that the required notice be increased to five years, to "provide greater political certainty" regarding the parties' intentions toward adherence. He notes that such an extension would not affect a party's right to immediately "suspend or terminate its obligations" in the event of a material breach by the other.[57]

This proposal has official echoes in both Moscow and Washington. The Soviet position would commit both sides to adhere to the Treaty for not less than ten years. The U.S. position offers adherence for not more than ten years, with deployment to follow.

The withdrawal clause issue then is one of demonstrating faith in the Treaty's value. But extending the timeline for withdrawal could have application beyond a near-zero BMD strategy. It could apply equally as well to a "virtual transition" scenario, indicating agreement to avoid premature, unilateral deployment of BMD while first-stage offensive force reductions were being completed.[58]

Deployment Issues

Deployment issues involve large phased-array radars (LPARs) and such dual-purpose, Treaty-exempted technologies as air defense and ATBM.

LPAR Deployments. The United States tried and failed to reach an arrangement in SALT I whereby LPARs could be deployed only by mutual agreement.[59] Today, the Soviet radars under construction at Krasnoyarsk and Baranovichi are of concern, the former because it violates Treaty siting provisions and the latter because of the "redundant" radar coverage it would provide.

As a solution to the LPAR issue, Thomas Longstreth and others would prohibit further deployment of LPARs, except as Treaty-permitted ABM radars or as early-warning radars on the national periphery and oriented outward. They would eliminate the Treaty's current exceptions for both space tracking and NTM purposes. Current LPAR deployments not in conformity with the new rules would be brought into conformity (that is, changed or dismantled).[60] This would require dismantling the Krasnoyarsk radar. According to some readings of the Treaty, it would also require halting construction of the LPAR at Fylingdales and dismantling the LPAR at Thule. It would not affect the Baranovichi construction, the siting of which is compliant for an early-warning radar.

Alternatively, all LPARs currently under construction could be "grandfathered," on the grounds that better early warning capability on both sides is good for crisis stability and that the risk of these radars contributing significantly to ABM battle management is a manageable one.

Another approach would trade Krasnoyarsk for Fylingdales; that is, Krasnoyarsk would be dismantled and Fylingdales' construction would be indefinitely postponed. Another may be to "grandfather" Thule and Krasnoyarsk, since both serve stabilizing early-warning functions, and to postpone indefinitely the construction of Baranovichi and Fylingdales. If Baranovichi's main purpose is redundant coverage of Pershing II flight corridors, that purpose disappears if Pershing IIs are withdrawn from Europe under an intermediate nuclear forces (INF) agreement. Fylingdales is the quid pro quo; an early-warning radar in Europe is exchanged for an early-warning radar in the western Soviet Union, neither of which has yet been built.

If these steps were taken and no further LPAR construction begun, both sides by the early 1990s still would have ringed their national territories with LPARs. If even those units clearly permitted by the Treaty eroded confidence of compliance with the Treaty's prohibition of a "base" for nationwide defense, then efforts to preserve the viability of the Treaty must pay greater attention to other elements of restraint, particularly intercept capability.

Air Defense and ATBM. The question here is whether verifiable distinctions can be drawn between increasingly capable SAM/ATBM systems and components, and BMD systems and components. If so, then they might be bilaterally negotiated, keeping ATBM unconstrained by the Treaty (up to agreed-upon performance limits), while better securing the Treaty from "upgrade" threats. If such distinctions are difficult to draw, then ATBM capabilities may need to be added to the technologies directly constrained by the Treaty.

Definitional Windows. Treaty constraints operate against systems to counter "strategic ballistic missiles." The first step in distinguishing ATBM from BMD is thus to ask: What is a strategic ballistic missile? For negotiating purposes, the Soviet Union often defines "strategic" in terms of systems that can hit the negotiators' homelands. By this definition, U.S. F-111s and other fighter bombers based in Europe are strategic weapons, as are carrier-based fighter bombers when the carriers sail within air attack range of the Soviet Union.

The United States has never accepted this definition. Since 1972, strategic systems by the U.S. definition are those limited by SALT: namely land-based ballistic missiles with ranges over 5,500 kilometers and ballistic missiles on nuclear submarines (with some variations to account for older Soviet missiles and boats).[61] The flight characteristics of the Soviet SS-20, whose 5,000 kilometer range lies between those of older SLBMs and current ICBMs, clearly would be similar to those of a strategic ballistic missile in SALT terms. A SALT-based definition of "strategic ballistic missile" would thus appear to forbid not only systems to counter SLBMs and ICBMs, but also systems to counter the SS-20, even though the SS-20 is not constrained by SALT. On the other hand, such a SALT-based definition would appear not to forbid systems to counter a Pershing II-class missile (1,800 kilometer range), which is certainly undesirable from a Western perspective.[62] By contrast, the Soviets' definition of "strategic" would forbid missile defenses to counter the Pershing II, but permit defenses to counter the SS-20, a technological and, from the Soviet viewpoint, a political absurdity. A compromise would encompass both these systems.

Several factors suggest 1,000 kilometers as a cutoff for the range of ballistic missiles against which defenses could be deployed without constraint by the Treaty. This is the range floor in the U.S.-Soviet negotiations on INF, the minimum range of missiles within the jurisdiction of the Soviet Strategic Rocket Forces, and just above the range from the inner-German border to Soviet territory. The longest range Soviet theater/tactical missiles are the SS-12 and SS-12mod (900 kilometers), the next being the SS-4 (2,000 kilometers and being retired), followed by the SS-20. Other than the Pershing II, the United States deploys only the Lance (110 kilometers).[63]

Defining this distinction between "strategic" and other ballistic missiles operationally and in a verifiable manner would be a logical next step.

Capability-Based Distinctions. Herbert Lin concludes that verifiable limits on ATBM systems might be established using two complementary measures: target detection range and interceptor rate of acceleration. These are the critical parameters of air defense/ATBM system performance. An observable surrogate measure for detection range in radars would be power-aperture product (average radiated power times antenna area, a measure already used in the Treaty). An observable surrogate for interceptor acceleration would be interceptor length and volume.[64]

Lin suggests a ceiling on power-aperture product for all mobile radars, based on the estimated product of the Patriot air defense radar (roughly 45,000 watts-meters-squared). Such a radar would detect a nominal strategic RV roughly 20 to 30 kilometers away, or about 5 to 8 seconds before RV impact. Without external cueing of the ATBM radar (e.g., with a longer-range infrared sensor like that aboard the Airborne Optical System, or a long-range radar like the LPAR at Fylingdales, or an LPAR sited in West Germany), incoming RVs might not be detected until they were well within that maximum radar range, too late to launch interceptors.[65]

The suggested length/volume limit on SAMs and ATBM interceptors would be 8 meters in length and 2.5 cubic meters in volume. Current-generation Soviet missiles (such as the SA-10 and SA-12) are under 7.5 meters long and under 2 cubic meters in volume. Current-generation NATO interceptors (HAWK and Patriot) are roughly 5 meters long and under 1 cubic meter in volume. By contrast, ABM interceptors are much larger in one or both dimensions. The U.S. Sprint is 8 meters long and about 4 cubic meters in volume (the Soviet SH-8 Gazelle, a similar missile, may be assumed to have similar dimensions). The U.S. Spartan long-range ABM interceptor is 17 meters long and 15 cubic meters in volume; and the Soviet Galosh ABM is roughly 20 meters long and 97 [*sic*] cubic meters in volume. Older U.S. and Soviet SAMs (the U.S. Nike Hercules and the Soviet SA-2, SA-4, and SA-5) exceed Lin's proposed limits. However, most of these are being phased out and in any case are less capable than the smaller missiles replacing them.[66]

Lin's proposed figures would "grandfather" current air defense missile systems. This would tend to simplify the negotiation of limits, but also would leave the SA-12 in the field. The operational SA-12A is apparently not the version with an ATBM capability; the SA-X-12B, evidently discernible by NTM, is the variant of concern.[67] The system's poor test record against tactical ballistic missiles indicates only marginal intercept capability against a Pershing II-class or longer-range (strategic) missile.

Lin suggests establishing testing limits to reinforce the performance limits just described. First, testing in an ABM mode would be defined as "the operation of one or more assemblies of parts observable by national technical means . . . in an exercise involving an ABM target. . . . An 'ABM target' would be defined as an object that achieves *either* a speed in excess of three kilometers per second or an altitude of greater than seventy kilometers."[68]

A weapons test "involving an ABM target" would include both attempts to intercept it and close, high-speed fly-bys of such a target (close but slow approaches, as in orbital docking maneuver, would be permitted).[69] Note that Lin's proposed limits tend to rule out weapons tests against objects in space. As formulated, however, they would tend not to rule out "point in space" tests.

Sixty to seventy kilometers is assessed to be the maximum altitude at which an interceptor with the estimated 100 kilometer slant range of the SA-12 could intercept an SS-12-class missile. Three kilometers per second is equivalent to the reentry velocity of an SS-12-class missile warhead.[70] These limits thus would leave both sides room to develop and deploy systems to counter aircraft and tactical ballistic missiles having ranges less than 1,000 kilometers, while they could increase confidence that ATBMs would not contribute significantly to strategic BMD. Agreed-upon ceilings on ATBM capabilities would also give the United States firmer grounds on which to challenge suspect Soviet weapons development.

As a retroactive measure, an agreement implementing such ceilings could ban the deployment of any non-ABM systems that were ever tested against "ABM targets" as defined by the new ceilings (e.g., the SA-X-12B, which is not yet operational and which would lose much of its purpose with the withdrawal of Pershing II and ground-launched cruise missiles from Europe).

However, some attention also may be needed regarding any "adjuncts" that would give an ATBM system much of its antimissile capability (airborne or spaceborne infrared sensors, and LPARs). An airborne infrared sensor supporting an ATBM might be something like the Airborne Optical System. If Western intelligence observed the Soviet Union testing an AOS—that had been developed as an adjunct to a BMD system—in concert with an ATBM radar and interceptor, then whatever compliance concerns either set of equipment generated on its own would be magnified many times. Soviet AOS would not be an adjunct in the context of ATBM, but rather an important component of its missile intercept capability. The airborne sensor may not give the ATBM "capabilities to counter" strategic ballistic missiles, but it would make the distinction finer and more difficult to verify.

By the same token, if extensively deployed SAM/ATBM units had access or opportunity to exercise with a large number of ABM radars scattered about the country, compliance problems would be significantly worse. A thin nationwide BMD system or a widely distributed terminal defense system could contribute significant detection and battle management capability to ATBM units linked to them and, at a minimum, result in a de facto terminal BMD system more capable than that agreed-upon bilaterally. Minimizing the ATBM problem thus requires continued contraints on deployment of actual ABM components.

Development Issues

Development issues include definitions of terms such as "component" and "development," which affect the Treaty's impact on predeployed ABM technologies and on non-ABM programs such as ASAT, under the aegis of which technologies having BMD potential might be tested in circumvention of Treaty restraints.

Definitional Issues. What constitutes an ABM component and what constitutes ABM component development (as distinguished from laboratory research and from non-ABM development) are key questions for the future efficacy of development restrictions in the Treaty.

Under the Reagan administration's current approach to development and testing, Treaty-constrained development begins "with the construction and testing of one or more prototypes of the system or its major components."[71] Earlier work is considered to be research and to be unconstrained. Nitze indirectly defines "component" in justifying the Delta 180 co-orbital intercept test of September 1986: "Since no device in the experiment was tested in an ABM mode and no device in the experiment could substitute for an ABM component, none of the devices in this experiment was itself an ABM component; the experiment, therefore, was fully consistent with the ABM Treaty."[72] This statement suggests two different standards for the definition of a "component": as something that is tested in an ABM mode, and as something that is capable of carrying out the function of, or substituting for, a traditional ABM component (a radar, launcher, or interceptor). The former standard offers better opportunity for effective constraints on testing of technologies that are not fixed, land-based, even technologies that could not be considered "prototype" ABM components, if coupled to an agreed-upon definition of ABM testing like that proposed by Lin.

Article VI(a), which contains the stricture "not to test . . . in an ABM mode," could be clarified as applying not only to non-ABM radars, launchers, and interceptors, but also to any other non-ABM sensors, launch platforms, and weapon technologies tested in an ABM mode. Relying on Article VI(a) to reinforce the development and testing limits of Article V would permit one side to call the other to account for efforts to track or intercept an "ABM target" with any "assembly of parts" that was not fixed, land-based, and sited at an ABM test range (or a Treaty-permitted ABM component at an ABM deployment site).

In applying standards of risk evaluation to such testing limits, one might acknowledge that some Soviet sensor tests against ABM targets might be missed by NTM, but significant tests very likely would not be missed at a greater rate than they are at present; the stringency of the limit and its bilaterally agreed-upon status would permit dubious activities to be questioned before the SCC at a fairly early stage. The net result would be a deeper development buffer against breakout, a desirable outcome if part of the goal is to strengthen the Treaty.

A more restrictive approach suggests that it would hurt the Treaty to tinker with technologies whose operational end-products could not legally be deployed (e.g., mobile, space-based components). To cope better with such technologies, Lin suggests redefining the constituent parts of an ABM system into a larger set of functions: target acquisition, tracking, discrimination (of targets from decoys), energy accumulation, intercept, kill, and kill assessment.[73] A technology capable of performing any one of these functions would be considered an ABM component for Treaty purposes. Defining components in this way would treat the parts of hybrid systems individually, such as a ground-based laser and a space-based mirror. The ground-based laser could be developed and tested under the Treaty, but the space-based mirror could not.

Refinements might take the form of negotiated, observable capabilities thresholds, the crossing of which would automatically categorize a technology as ABM-capable and therefore ban it in other than a fixed-base mode, to be tested only at ABM test ranges. Threshold brightness levels and mirror surface areas for lasers, threshold power levels for particle beams, and firing rates for railguns all have been suggested.[74] Compliance with threshold limits would be easier to monitor if the verification process did not have to make fine distinctions (e.g., between a BMD laser and an ASAT laser, or between a neutral particle beam [NPB] generator for killing ballistic missile boosters and one for interactive midcourse discrimination of warheads from decoys). As noted earlier, the difference in brightness requirements for a beam weapon and for such non-ABM applications as laser communications or laser radars is sufficiently great (more than 10 billionfold) that thresholds suitable for limiting weapons development would readily accommodate most non-ABM, non-weapon applications of directed energy devices.[75]

The threshold capability approach would tend to halt development of a technology or device well before it became capable of substituting for an ABM component. Sufficiently low capabilities thresholds would discourage work-around strategies, and order-of-magnitude uncertainties about a device's power or brightness need not threaten the integrity of the constraints (although levels thought to cross a given threshold should be subject to immediate challenge before the SCC or some like body).

Distinctions between unconstrained laboratory research and constrained development and testing could still be maintained by recourse to the verifiability criterion in Ambassador Gerard Smith's 1972 definition of development.[76] What could be seen by NTM as exceeding agreed-upon performance thresholds would be cause for a compliance query before the SCC. However, the regime might allow for exceptions cast in terms of specific applications that used otherwise prohibited technology (the Soviets, for example, will be sending a high-energy laser to Mars aboard their Phobos probe in mid-1988); or exceptions could be cast in terms of test quotas for non-fixed-land-based devices; or in terms of the numbers of such devices present at any one time aboard orbital "laboratories."[77] Exceptions for scientific experiments could be subject to joint, on-site inspection. Other excepted tests would be monitored by NTM (perhaps a monitoring satellite that would travel in a parallel orbit).

The key difference between this arrangement and current practice would be its basis in bilateral agreement. The United States and the Soviet Union might agree to allow such limited space laboratory testing as part of a package agreement to shore up the Treaty, recognizing it as an element of both sides' programs to keep up with the state of the art in BMD and to avoid technological surprises. A case can be made that it would be better to allow such research under agreed-upon and controlled conditions than either to attempt to stifle it completely or to conduct it on the basis of unilateral definitions of terms that may only rationalize a creep-out strategy.

On the other hand, research in exotic technologies with BMD potential may still be doable if confined to fixed, land-based test sites. Such tight constraints on weapons testing would be especially important if testing of space-based and other mobile sensors were judged difficult to limit verifiably. Moreover, such sensors can and do perform desirable functions (e.g., early warning), and those in low earth orbit could have a variety of non-BMD applications (e.g., aircraft tracking and ocean-vessel surveillance). Once again, this suggests that primary attention must focus on limiting launch platforms and kill mechanisms.

Weapons Test Limits. The suggested criteria for testing in an ABM mode could be implemented in one of three ways. The first would adapt the criteria to ban exo-atmospheric testing in an ABM mode; that is, all weapons tests, including those of ABM interceptors and "point in space" tests, would be confined to perhaps 70 kilometers altitude, permitting tests of Sprint-class terminal interceptors that are suited to point defense, but not longer range interceptors that are suited to area defense. ASAT tests would be "in an ABM mode," since their targets would exceed 70 kilometers altitude and 3 kilometers per second. Thus they would be banned as well—although current U.S. and Soviet weapons might be excepted, since their ABM potential is nil. The limit would prevent deployment of the sort of accident-absorbing ABM system discussed earlier, but would also prevent deployment of ASATs capable of posing serious threats to U.S. satellites.

Due to their inherent exo-atmospheric capability, testing of directed energy devices exceeding agreed-upon brightness thresholds, including fixed, land-based devices, would be banned under this first implementation.

All testing "in an ABM mode" would be confined to agreed-upon ABM test sites and to fixed basing modes. This would rule out the testing of such technologies as the Airborne Optical Adjunct, since its primary function is tracking strategic missile RVs before reentry and it is not a fixed, land-based system.

The second implementation option would permit testing in an ABM mode of devices that exceeded agreed-upon thresholds, but only in fixed, land-based modes and only at agreed-upon ABM test areas, whatever the purpose of the test (ABM, ASAT, or other).[78] Any space-directed weapons testing would be limited to such test areas. Logically, however, to prevent erosion of ABM Treaty limits, the deployment of ASAT capability beyond that provided by the current U.S. miniature homing vehicle system and the Soviet co-orbital ASAT system, and by permitted deployments of 100 ABM interceptors, should also be banned.

The third implementation option would modify the second to allow for the sorts of excepted space experiments discussed above.

Under all three implementation options, the development, testing, and deployment of SAMs and ATBMs that did not exceed agreed-upon component performance and testing thresholds would be permitted.

Summary

The foregoing discussion suggests that, given sufficient interest and political will on both sides, measures could be negotiated that would protect the ABM Treaty from threats from below (SAMs and ATBMs), while preserving rights to test and deploy such systems for tactical purposes and for defense against cruise missiles and aircraft. However, equally as important as direct ceilings on such systems' intrinsic performance would be constraints on other components that functioned as force multipliers for them.

Limits on development and testing designed to enforce a deeper buffer against high-technology erosion of the Treaty could usefully focus on the concept of "testing in an ABM mode," backed by low performance thresholds for exotic technologies, rather than force NTM to make fine distinctions between technologies that may or may not be capable of substituting for an ABM component. Such an approach could reduce disputes within the U.S. government about Soviet compliance behavior, as well as bilateral disputes before the SCC or in the public arena. This approach may increase some monitoring uncertainties, but it arguably would not pose any increased risk to national security.

Finally, current radar deployment disputes could be resolved in ways that left both sides with enhanced early-warning capability, ways that removed questionable radars, or ways that removed radars now permitted under the Treaty but that are cause for concern.

NOTES

1. As ACDA Director Adelman observes, but for the broad interpretation "we would have to abrogate the treaty to do the testing we need to see if an SDI works, and no president is going to abrogate the SDI treaty [sic] just to do testing." *Washington Post,* 10 Feb. 1987, p. 11.
2. Stephen M. Meyer, "The U.S. SDI and Soviet Defense Policy: Near-Term Impact and Responses" (Paper delivered to the Aspen Arms Control Workshop, Aspen, Colo., July 1986). Cited with permission.
3. Gen. Hoyt C. Vandenberg press conference, reported in *NYT,* 18 Jan. 1951, p. 4. Vandenberg said that 20 to 30 percent attrition would be "extraordinary." On U.S. SAM effectiveness, see *NYT,* 21 Nov. 1955, p. 6; idem, 21 May 1956, p. 1; idem, 23 May 1956, p. 1; idem, 24 May 1956, p. 8; and idem, 24–25 May, 1959, p. 1. These tests appear not to have involved U.S. Air Force fighter interceptors, just Army SAMs, and it was in SAC's interest to demonstrate that "the bomber will always get through." Still, even an eight- or ninefold increase in defense effectiveness would not have saved the cities under attack.
4. See Ashton Carter, "Directed Energy Ballistic Missile Defense in Space," Background paper prepared for the U.S. Congress (Washington, D.C.: Office of Technology Assessment, April 1984), pp. 19–20. Carter's report was attacked by the Office of the Secretary of Defense and by analysts at Los Alamos National Laboratory for containing "technical inaccuracies." However, an independent OTA review panel confirmed Carter's analyses. See "Pentagon's Defensive Criticism Turned," *Nature* (August 1984): 353.

See also Boyce Rensberger, "How Many Satellites?" *Washington Post*, 4 Mar. 1985; and, for a technical assessment, see Richard L. Garwin, "How Many Orbiting Lasers for Boost-Phase Intercept?" *Nature* (23 May 1985): 286–90.

5. The probability that a warhead would survive passage through three successive, independent layers of 75 percent effective defense would be $(1 - SSPK_d)^n$, where $SSPK_d$ is the probability that the defense will destroy a warhead with one shot, and "n" is the number of defense layers (or, alternatively, the total number of shots made at the warhead). This is a defense-optimistic calculation with no entropic factors (such as unfriendly nuclear detonations) interfering with the sequence of intercept attempts.

6. Peter D. Zimmerman, "Pork Bellies and SDI," *Foreign Policy* 63 (Summer 1986): 76–87. In their argument for a transition to defenses, Payne and Gray consider the prospect of an untested defense failing catastrophically to be a bonus for stability in a defended world, since a would-be aggressor could never be certain that its defenses would work. However, by this logic, and as noted in Chapter 1, a would-be aggressor would be more deterred if it had no defenses at all, all else being equal. See Keith Payne and Colin Gray, "Nuclear Policy and the Defensive Transition," *Foreign Affairs* (Spring 1984): 827.

7. Carter, "Directed Energy," note 4, pp. 68–69.

8. Zimmerman, "Pork Bellies and SDI," note 6; Carter, "Directed Energy," note 4, pp. 65–69; Kevin Lewis, "BMD and U.S. Limited Strategic Employment Policy," *Journal of Strategic Studies* (Summer 1985): 125–44.

9. On current Soviet weapon inventories, see IISS, *The Military Balance, 1985–86* (London: IISS, 1985), pp. 180–81. Equivalent megatonnage is defined as weapons yield taken to the two-thirds power. Arthur Katz cites a 1970 Stanford Research Institute study's conclusions that 500 1-megaton nuclear weapons would "neutraliz[e] the productive capacity of the 71 [largest] U.S. standard metropolitan statistical areas," from New York through Fresno. See Arthur Katz, *Life After Nuclear War* (Cambridge, Mass.: Ballinger Publishing Company, 1982), pp. 95–97. The SRI study cited is Richard Goen, R. Bothun, and F. Walker, *Potential Vulnerability Affecting National Survival* (Menlo Park, Calif.: Stanford Research Institute, September 1970), p. 31.

10. Stephen M. Meyer, "Soviet Strategic Programs and the U.S. SDI," *Survival* (November/December 1985), pp. 289–90.

11. See U.S., Department of State, *The Strategic Defense Initiative*, Special Rept. 129 (Washington, D.C.: U.S. Government Printing Office, June 1985), pp. 2, 5; also Payne and Gray, "Nuclear Policy," note 6, pp. 823–24; and Colin Gray, "Deterrence, Arms Control, and the Defense Transition," *Orbis* (Summer 1984): 229–30. On U.S. determination to react to Soviet BMD, see Caspar Weinberger, "Responding to Soviet Violations

Policy (RSVP) Study," Memorandum for the President, November 1985, p. 10: "Even a *probable* territorial defense would require us to increase the number of our offensive forces and their ability to penetrate Soviet defenses to assure that our operational plans could be executed. [Emphasis in original] "

12. See, for example, Payne and Gray, "Nuclear Policy," note 6. Recent RAND Corporation analyses suggest that the United States would need to deploy defenses as much as two times faster than the Soviet Union in a competitive transition scenario in order to maintain crisis stability. See Glenn A. Kent and Randall J. DeValk, *Strategic Defenses and the Transition to Assured Survival*, RAND Project Air Force Report R-3369-AF. (Washington, D.C., October 1986); and Dean A. Wilkening, et al., "Strategic Defenses and First-Strike Stability," *Survival* (March/April 1987): esp. p. 147. (Article based on RAND Report R-3412-FF/RC.) See also JASON/MITRE, *Deployment Stability of Strategic Defense*, Report JSR-85-926. (McLean, Va.: MITRE, Oct. 1986), esp. pp. 2-13.

13. See, for example, Colin Gray, "Nuclear Delusions, Six Arms Control Fallacies," *Policy Review* (Summer 1986): 51; and *idem*, "The Transition from Offense to Defense," *Washington Quarterly* (Summer 1986), pp. 68, 71. For statements by Perle, see Adam Pertman, "Pentagon Aide Challenges Congress," *Boston Globe*, 6 June 1986, p. 9.

14. For a historical overview of Russian/Soviet behavior while under external pressure, see Joseph Whelan, *Soviet Diplomacy and Negotiating Behavior*, issued by the Committee on Foreign Affairs, U.S. House of Representatives, Special Studies Series on Foreign Affairs Issues, Vol. 1, 1979. For recent U.S. intelligence community assessments of Soviet defense spending trends and prospects, see testimony of Douglas MacEachin and Robert Schmitt in U.S., Congress, Joint Economic Committee *Allocation of Resources in the Soviet Union and China—1985*, Pt. 11 99th Cong., 2d sess., March 1986, esp. pp. 19, 110, 119.

Although the Soviet Union is expected to rise to meet any U.S. challenge, arms control agreements that "included sizeable reductions in strategic forces and prevented or delayed deployment of U.S. SDI programs, would provide substantial economic benefits to the USSR." Ibid., p. 9.

The rate of increase in Soviet military spending has slowed in the past decade as the Soviet Union's overall economic growth rate has slowed. Total military investment is now thought to have increased at roughly 1 percent annually from 1975 through at least 1981, having increased at much higher rates from 1965 through 1974. The CIA, which includes all military procurements in its estimates, shows a steady amount of procurement from 1982 through 1984. The DIA, which calculates the procurement costs of only major weapon systems, shows an increase of 3 to 4 percent annually, from 1982 through 1984. However, the spending base for the Soviet military is quite substantial, as a legacy of its growth in the first decade of Brezhnev's rule. Thus, several sectors of the military have been extensively modernized during the period of slow growth, including the strategic rocket forces, tank forces, theater missile forces, and tactical aviation. Ibid., pp. 3, 36.

15. Congress authorized $292 billion in defense spending for fiscal 1987, down 2+ percent in real terms for the second year in a row, after inflation-adjusted increases of 11.8, 8.0, 4.8, and 7.2 percent in fiscal years 1982 through 1985. Caspar Weinberger, *Annual Report to the Congress, Fiscal Year 1987* (Washington, D.C.: Office of the Secretary of Defense, February 1986), p. 313 (budget tables); and Richard Halloran, "Weinberger is Determined on Arms Budget," *NYT*, 7 Nov. 1986, p. A15.

SDI has been assigned the highest program priority by the Defense Department and has been exempted from internal departmental budget cuts mandated by the Gramm-Rudman-Hollings deficit reduction legislation, but it remains a visible fiscal target of Congress nonetheless. Rather than generating the steady support that advanced technology research needs, SDI may fall prey to the traditional boom-and-bust defense funding cycle, a victim in part of its own hard sell. Bill Keller, "Missile Shield Program Gets Pentagon's Highest Priority," *NYT*, 29 Jan. 1986, p. A11; Jonathan Fuergringer, "House . . . Decides to Reduce 'Star Wars' Money," *NYT*, 13 Aug. 1986, p. 1.

16. Dean A. Wilkening, "Space Based Weapons," in William Durch, ed., *National Interests and the Military Use of Space* (Cambridge, Mass.: Ballinger Publishing Company, 1984), pp. 155–60, 163. Space-basing would routinely bring opposing systems within range of one another, since major elements of the two systems both would be placed in low earth orbit.

17. Kent and DeValk, "Strategic Defenses," note 12, p. 2; Wilkening, et al., "Strategic Defenses," note 12, pp. 145–46. Passive and active satellite defense measures (DSAT) could account for a significant fraction of the system mass lofted into orbit. To deal with the space- and ground-based ASAT threats, each space component cluster may come to resemble a sea-going convoy or carrier battle group, most of the firepower of which is devoted to self-protection. However, in the case of naval forces, only three battle groups are generally required to keep one on station continuously; only one in ten space-based battle groups would be on station.

18. Charles Glaser, "Why Even Good Defenses May Be Bad," *International Security* (Fall 1984): 121.

19. I am grateful to Charles Glaser for emphasizing the appropriateness of EMT as the measure of merit in this case.

20. Gray, "The Transition," note 13, pp. 62–63, 66–67.

21. As of mid-1987, it looked as though the spirit of agreement, if any, might flow in the opposite direction—from shorter range missile reductions to the strategic arms talks.

22. Glenn Kent, "A Coherent Package—Arms Control, Zero Ballistic Missiles, and Strategic Defense," transcript of testimony before the U.S. Congress, House Armed Services Committee, 26 Mar. 1987, p. 10. See also Glenn Kent, "A Suggested Policy Framework for Strategic Defenses," RAND Note N-2432-FF/RC, December 1986.

23. Kent and DeValk, *Strategic Defenses*, note 12, p. 14; Kent, "A Coherent Package," note 22, pp. 6–7, 12. Kent does not suggest eliminating bombers because he believes the West needs to maintain an offensive nuclear deterrent vis-à-vis the Soviet Union. Since he sees no "worthy goals" for BMD other than protecting a missile elimination agreement, whether the United States should pursue such an elimination agreement is, in Kent's view, the primary policy issue.

24. Kent, "A Coherent Package," note 22, p. 11.

25. Although the need to deploy some BMD and the overall size of the deployment might be agreed-upon at the outset as part of the total arms control package, negotiation of the final go-ahead for BMD and the details of deployment might wait until missile force reductions were well underway and had generated a certain amount of confidence in dismantlement and verification procedures.

26. To support development and deployment of such a system would require changes to Article I's prohibition of a base for territorial defense, to Article III's limits on BMD deployment (numbers and areas), and perhaps to Article V's limits on development, testing, and deployment of mobile components if, for example, space-based sensors permitted an anti-cheating system to function with fewer deployed interceptors. Survivability and stability concerns would militate against space-based weapons, however. Even kinetic BMD weapons in orbit could double as global ASATs for the region of low earth orbit, serving both as defense-suppression devices and as targets.

27. Thomas Schelling, *Arms and Influence* (New Haven: Yale University Press, 1966), pp. 248 ff.

28. Kent, "A Coherent Package," note 22, pp. 11–12.

29. This section does not discuss defense of C^3, deferring to Carter's assessment that, although any Soviet attack against U.S. C^3 is likely to be massive in nature, "It is almost impossible to erect an analytic case for the proposition that current U.S. strategic forces can be literally decapitated by a Soviet strike." On the other hand, highly damaging strikes on C^3 are unlikely in limited war, "which [would] by definition involve very much smaller and more discriminating use of nuclear weapons, [and] also by definition spare the strategic command system." See Ashton Carter, "Assessing Command System Vulnerability," in Ashton Carter, John Steinbruner, and Charles Zraket, eds., *Managing Nuclear Operations* (Washington, D.C.: Brookings Institution, 1987), pp. 606–7. For tables of C^3 targets, see pp. 561–62.

Thus, BMD is not needed to ensure initial retaliation in a full-scale war, nor in limited war. BMD that might limit damage to the far-flung C^3 system in a full-scale war would, moreover, amount to a full-scale nationwide defense and not to a limited BMD deployment.

30. Caspar Weinberger, *Annual Report to the Congress for Fiscal Year 1986* (Washington, D.C.: Office of the Secretary of Defense, February 1985), p. 27. ("Should deterrence fail, U.S. strategy seeks the earliest termination of conflict on terms favorable to the United States. . . . 'Favorable' means that if war is forced upon us, we must win. . . . [T]he United States cannot prepare only for a 'short war,' which would merely tempt an adversary to believe he could outlast us in combat.") Richard Halloran, "Pentagon Draws Up First Strategy for Fighting a Long Nuclear War," *NYT*, 30 May 1982, p. 1; Weinberger clarification in letter to the editor, *Los Angeles Times*, 25 Aug. 1982.

31. *Report of the President's Commission on Strategic Forces*, April 1983, pp. 7–8. See also U.S. Office of Technology Assessment, *MX Missile Basing* (Washington, D.C.: U.S. Government Printing Office, 1981), pp. 303–6, for a summary of ICBM characteristics.

32. Paul Nitze, "Deterring Our Deterrent," *Foreign Policy* 25 (Winter 1976–77); 195–210; idem, "Assuring Strategic Stability in an Era of Détente," *Foreign Affairs* 54:2 (January 1976): 207–32; and his testimony in U.S., Congress, Senate, Foreign Relations Committee, *Hearings on SALT II Ratification*, 96th Cong., 1st sess., 1979, p. 450.

33. Albert Carnesale and Charles Glaser, "ICBM Vulnerability: The Cures Are Worse Than the Disease," *International Security* (Summer 1982): 76–78. In the above-cited *Foreign Affairs* article, Nitze does acknowledge that the "absolute level of forces remaining to the weaker side" would be an important factor in intrawar deterrence. If high enough "to threaten a major portion of the other side's military and urban/industrial targets" and "under effective command and control," it would be "conducive to continued effective deterrence even if the ratios are unfavorable" (p. 226). Meyer notes that relative measures of capability *would* enter into pre-war Soviet damage-limitation calculations and could influence a decision to preempt "if Soviet leaders had great confidence in the reliability of their strategic warning and believed nuclear war was unavoidable." See Stephen Meyer, "Soviet Perspectives on the Paths to Nuclear War," in Allison, Carnesale, and Nye, eds., *Hawks, Doves and Owls* (New York: W.W. Norton, 1985), p. 200.

However, this need not imply that U.S. ICBMs are a sufficiently "attractive nuisance" to Soviet decisionmakers that a decision to attack might be made *independent* of warning that the West was about to launch its forces.

34. Carnesale and Glaser, "ICBM Vulnerability," note 33, pp. 79–80.

35. In 1977 Congressional testimony on this subject, Secretary Brown estimated that it would take twelve hours to reload silos, provided extra mis-

siles were stored at the launch site. (The lapsed SALT II agreement prohibited such on-site storage.) See U.S., Congress, House, Armed Services Committee, *Hearings on H.R. 8390 and Review of the State of U.S. Strategic Forces*, 95th Cong., 1st sess., 1977, p. 208. In 1980, *Aviation Week* (22 September 1980: 14–15) reported Soviet reload exercises that took two to five days. Both items are cited by Desmond Ball, *Targeting for Strategic Deterrence*, Adelphi Paper No. 185 (London: IISS, 1983), Ch. 3, n. 10.

In early 1981, the Joint Chiefs of Staff estimated that "the Soviets probably cannot refurbish and reload silo launchers in less than several days." See Organization of the Joint Chiefs of Staff, *United States Military Posture for 1982* (Washington, D.C.: Organization of the Joint Chiefs of Staff, February 1981), p. 100. Testifying before the House Appropriations Committee in 1982, Air Force General Kelly Burke estimated "two or three days as about a minimum time" for the Soviets to refurbish and reload SS-17 and SS-18 silos. U.S., Congress, House, Committee on Appropriations, *Department of Defense Appropriations for Fiscal Year 1983, Part 4*, 97th Cong., 2d sess., 20 Apr. 1982, p. 551.

More recently, Defense Department statements on Soviet ICBM reload capabilities have stressed the capability, but omitted estimates of time required. See, for example, *SMP*, 1984 and 1985 eds. (Washington, D.C.: U.S. Government Printing Office), pp. 21 and 28, respectively.

SS-17s are based at two deployment areas in the northwestern Soviet Union; and SS-18s are based at six deployment areas in the south central Soviet Union. *SMP*, 1984, p. 21. Note that although the Soviet Union has evinced little interest in silo defense per se, as discussed in text, defenses against second-strike bomber and cruise missile counter-silo attacks could be augmented by BMD designed to prevent SLBM pin-down of silo reloading.

36. Carnesale and Glaser, "ICBM Vulnerability," note 33, pp. 80–82.

37. John Steinbruner, "Nuclear Decapitation," *Foreign Policy* 45 (Winter 1981–82); Stephen M. Meyer, "Soviet Nuclear Operations," in Carter, et al., *Managing Nuclear Operations*, note 29, pp. 529–31.

38. Harold Brown, "Is SDI Technically Feasible?" *Foreign Affairs* (Winter 1985–86): 436–37.

39. Charles Glaser, "Do We Want the Missile Defenses We Can Build?" *International Security* (Summer 1985): 30–32.

40. David Holloway, *Soviet Union and the Arms Race* (New Haven: Yale University Press, 1983), Ch. 3, esp. pp. 57–8; also Meyer, "Soviet Nuclear Operations," note 37.

41. Ibid. See also Vann H. Van Diepen, "Strategic Force Survivability and the Soviet Union" (M.A. thesis, Department of Political Science, MIT, February 1983).

In their assessment of a Soviet nuclear exchange model that measures second-strike strategic reserves, Meyer and Almquist note that, because the model emphasizes deliverable second-strike EMT, the more survivable Soviet forces are, the better the model's results from a Soviet perspective. Peter B. Almquist and Stephen M. Meyer, *Insights from Mathematical Modeling in Soviet Mission Analysis (Part II)*. Report prepared for the U.S. Department of Defense Advanced Research Projects Agency (Cambridge, Mass.: Department of Political Science and Center for International Studies, MIT, 1985), p. 24.

42. Sayre Stevens, "The Soviet BMD Program," in Carter and Schwartz, eds., *Ballistic Missile Defense* (Washington, D.C.: Brookings Institution, 1984), pp. 219–20.

43. William Durch and Peter Almquist, "The East-West Military Balance," in Blechman and Luttwak, eds., *International Security Yearbook, 1984–85* (Boulder, Colo.: Westview Press, 1985), pp. 25–27.

44. Kent and DeValk, *Strategic Defenses*, note 12, pp. 56–57. "Discriminating" defenses can distinguish silo-killing warheads from others in the attack; "semipreferential" defenses are pre-allocated to defend some fraction of silos more vigorously than others, and allocations do not change when the attack begins. A completely preferential defense would allocate its interceptors in response to the attack, but a local, endo-atmospheric defense would not have time to do so. The kill probabilities are simplifying assumptions that are optimistic for the offense (Soviet SS-18 warheads may have a single-shot kill probability of about 0.7 against U.S. silos) and also highly optimistic for the defense. Thus the equations return the *minimum* number of interceptors required to enforce a given level of silo survival against an attack of a given size.

   Should strategic offensive forces be reduced by 50 percent across the board, the numbers of interceptors required to enforce 50 percent survival of the remaining silo-based ICBMs would scale proportionately to about 2,000 to 3,600, and the consequences for U.S. retaliation would be similar.

45. Kent, "A Coherent Package," note 22, pp. 8–9.

46. On Sprint capabilities, see R.T. Pretty, ed., *Jane's Weapon Systems, 1976* (New York: Franklin Watts Press, 1976), pp. 65, 590. The intercept range of the Soviet SH-08 Gazelle would be similar. Lacking deception capability or mobility, missile defense components would be prime targets for the offense and, depending on the RV/target ratio of reduced strategic forces, could be at risk of saturation. Permitting rapid reload of fixed launchers of endo-atmospheric interceptors would allow deployment of a more capable fixed defense, without expanding its footprint but, like mobility, this also would make estimating the other side's actual BMD capability very difficult. This is not bad for deterrent stability, since the system is a point defense, but it is not good for the stability of an arms control regime that is supposed to place a ceiling on the level of deployed BMD.

47. The focus here is on hard-target-capable ballistic missile warheads because the defense of the ICBM force is intended in large part to preserve or enhance U.S. prompt hard-target-kill capability. (Hard-target-capable is defined somewhat arbitrarily as a kill probability of 0.90, given two-on-one targeting against a target hardened to withstand 1,000 pounds per square inch static overpressure.) Assume a generated alert with the following U.S. missile forces: 1,000 ICBMs in silos (500 Minuteman III, 50 MX, and 450 Minuteman II—the latter absorb Soviet RVs but are not included in the retaliatory calculations because they are not hard-target-capable); 500 Midgetman mobile ICBMs (not defended); and 336 Trident II (D5) SLBMs aboard 14 submarines. Alert rates are assumed to be 90 percent for all of the ICBMs and 85 percent for the SLBMs (due to surge from port on generated alert). Overall reliability figures for alert forces are assumed to be 90 percent for the silo-based ICBMs, 85 percent for the mobile ICBMs, and 80 percent for the SLBMs. These assumptions yield about 2,000 alert and reliable ICBM warheads and about 1,825 alert and reliable Trident II warheads. Half of the defended, silo-based warheads (roughly 800) survive attack, as do all of the alert SLBM warheads and 80 percent (or roughly 300) of the mobile ICBMs (it is assumed that a properly dispersed force would be difficult to barrage, but that the force would nonetheless suffer some attrition in an attack). In round numbers, about 2,900 hard-target-capable ballistic missile warheads are available to retaliate.

The formula used to calculate the probability of warhead penetration through the defense is $(1 - SSPK_d)^n$, where $SSPK_d$ is the defense's single-shot kill probability and "n" is the number of shots taken at an incoming warhead. If 8,000 Soviet ABM interceptors have even a 0.5 single-shot PK, then assuming that three interceptors are used against each of 2,200 incoming RVs and two against the remaining 700 RVs, only 450 U.S. RVs would reach their targets.

In the case of near-zero BMD, 10 percent of the silo-based ICBM force, or about 160 warheads, are assumed to survive and to be available to retaliate, joined by the 300 Midgetman and the 1,825 Trident II warheads, or about 2,300 retaliatory warheads, in round numbers, all of which penetrate to their targets.

Alert rates, reliability figures, and force size estimates for submarines are adapted from U.S., Congressional Budget Office, *Retaliatory Issues for the U.S. Strategic Nuclear Forces*, Background Paper (Washington, D.C.: U.S. Government Printing Office, June 1978), pp. 9, 12; and idem, *Modernizing the U.S. Strategic Offensive Forces: The Administration's Program and Alternatives*, Study (May 1983), pp. 82, 84. Blair notes that "[T]he Navy is generally credited with the capability to surge out to sea the entire SSBN fleet, except for boats in dry dock, in 24 to 48 hours." Bruce Blair,

"Alerting in Crisis and Conventional War," in Carter, et al., eds., *Managing Nuclear Operations*, note 29, p. 89(n).

See also U.S., Congressional Office of Technology Assessment, *Ballistic Missile Defense Technologies* (Washington, D.C.: U.S. Government Printing Office, September 1985), pp. 104–106, for additional discussion of the relative impact of BMD for ICBM silos on relative U.S. retaliatory capability.

For a discussion of Soviet preferences in BMD deployments, see Sayre Stevens, "The Soviet BMD Program," in Ashton Carter and David Schwartz, eds., *Ballistic Missile Defense* (Washington, D.C.: Brookings Institution, 1984), pp. 219–20.

48. Kent and DeValk's perfect ABM interceptors, if sufficiently plentiful, could also protect such soft targets as SAC bases. However, more realistic performance assumptions make airbase defense problematic. One "leaker" with the yield of an average Soviet strategic warhead could destroy a base and its forces if detonated anywhere within several miles of the base.

49. *Strategic Survey, 1987* (London: IISS, 1987), pp. 41–42.

50. SS-18 deployments are described in Desmond Ball, "Research Note: Soviet ICBM Deployment," in *Survival* (July/August 1980). See also William Arkin and Richard Fieldhouse, *Nuclear Battlefields* (Cambridge, Mass.: Ballinger Publishing Company, 1985), Appendix B. Numbers of SS-11 silos at Gladkaya are a rough estimate derived by averaging the total deployment of 450 SS-11s over five dedicated and two shared deployment areas.

51. Paul Bracken, "Accidental Nuclear War," in Allison, et al., *Hawks, Doves and Owls*, note 33, p. 38.

52. Ibid., p. 39.

53. Ibid., pp. 40–45; and Cotter, "Peacetime Operations," note 29, pp. 42–52.

54. Meyer, "Soviet Nuclear Operations," note 37, pp. 491–93.

55. U.S. Arms Control and Disarmament Agency, *Arms Control and Disarmament Agreements, 1980 Edition* (Washington, D.C.: U.S. ACDA, 1980), pp. 109–12.

56. IISS, *The Military Balance, 1986–87* (London: IISS, 1986), pp. 140–41.

57. John B. Rhinelander, "U.S. and Soviet Ballistic Missile Defense Programmes: Implications for the 1972 ABM Treaty," *Space Policy* (May 1986): 149.

58. The U.S. proposal was tied to a two phase reduction in offensive arms and could be consistent with a "virtual transition" scenario, but the Reykjavik proposal seems to contemplate full-scale SDI employment and not a minimum anti-cheating system.

59. Gerard Smith, *Doubletalk* (Garden City, N.Y.: Doubleday, 1980), p. 313.

60. Thomas K. Longstreth, John E. Pike, and John B. Rhinelander, *The Impact of U.S. and Soviet Ballistic Missile Defense Programs on the ABM*

Treaty (Report prepared for the National Campaign to Save the ABM Treaty, Washington, D.C., March 1985), p. 68.

61. U.S., Department of State, *SALT II Agreement,* Selected Documents No. 12 (Washington, D.C.: U.S. Government Printing Office, June 1979), pp. 3–5.

62. Although the range of the Pershing II exceeds that of the Soviet SSN-5 SLBM (1,800 versus roughly 1,000 kilometers) and some SSN-5s were deployed on SALT-accountable Hotel-class SSBNs in 1972, when SALT I was signed, SSN-5s are now deployed only on diesel-powered Golf-class boats and are not SALT-accountable. This argues that the minimum range of a SALT-defined strategic missile is now that of the Soviet SSN-6 (2,400 kilometers). (The shortest range U.S. SLBM deployed in 1972 was the Polaris A2, with a range of roughly 2,800 kilometers.) See Theodore Postol, "The Trident and Strategic Stability," *Oceanus* (Summer 1985): 45; and IISS, *The Military Balance, 1986–87* (London: IISS, 1986, p. 205.

63. Ibid., pp. 200–205. West Germany deploys 72 Pershing IAs (range 720 kilometers). France deploys 18 SSBS-3 missiles (3,500 kilometers). French SLBMs have a range of 3,000 kilometers or greater, and British SLBMs a range of 4,600 kilometers.

64. Herbert Lin, "New Weapon Technologies and the ABM Treaty" (Paper prepared for Defense and Arms Control Studies Program, Center for International Studies, MIT, 1 June 1987, mimeographed), Appendices I and II. Radar detection range varies not only with power-aperture product, but with radar cross section of the target and the solid angle through which the radar must search for the target. Targets with larger radar cross sections (e.g., aircraft) are detected further out. External cueing that tells the radar where to look for targets can reduce the search volume and permit the radar to acquire the target sooner and thus further out. Morel and Postol emphasize that simple design changes can make shorter range missiles and RVs as hard to detect as longer range RVs, especially when nose-on to the ATBM radar (i.e., in a defense suppression attack). Decoys also exact relatively less of a penalty in shorter range systems, since a larger fraction of their total mass is payload. Higher angles of reentry mean there is less opportunity for atmospheric friction to separate decoys from RVs, and radar jammers can themselves be loaded onto ballistic missiles and fired ahead of the main attack to blind defense radars. Benoit Morel and Theodore Postol, "ATBM Technologies and NATO," in Jeffrey Boutwell and Donald Hafner, eds., *European Missile Defense: ATBMs and Western Security* (Cambridge, Mass.: Ballinger Publishing Company, forthcoming).

Interceptor volume is a useful measure because an interceptor with a given volume (and thus a given amount of fuel) can be designed to burn its fuel slowly and fly a long distance, or to burn it fast and reach a nearer intercept point very quickly. To increase the range of a fast-burn interceptor, its fuel load (and hence its volume) must increase. Length is a good surrogate for volume because the diameter of an interceptor can only grow so large before atmospheric drag cuts into its performance (there is an optimal cone shape for hyper-acceleration missiles). An interceptor limited to a given length thus will be limited to a knowable speed/range envelope as well. However, improvements in guidance and other technologies over time may allow a greater fraction of mass to be devoted to fuel, so that limits may need to be lowered from time to time to remain viable constraints on BMD capability.

65. Ibid.; and Lin, "New Weapon Technologies," note 64, p. 71.
66. Ibid., pp. 68–70.
67. SMP, 1987, p. 50; Longstreth, et al., Impact, note 60, pp. 55–56.
68. Lin, "New Weapon Technologies," note 64, pp. 4–5.
69. Ibid., p. 60.
70. Ibid., pp. 57–58.
71. Paul Nitze, "Permitted and Prohibited Activities Under the ABM Treaty," Current Policy Rept. 886 (Washington, D.C.: U.S. Department of State, Bureau of Public Affairs, November 1986), p. 2.
72. Ibid.
73. Lin, "New Weapon Technologies," note 64, pp. 79–80.
74. Ibid., pp. 84–90; also John B. Rhinelander, "Specific Proposals to Strengthen the ABM Treaty" (Paper prepared for Conference of U.S. and Soviet Legal Experts, Sponsored by the Lawyers Alliance for Nuclear Arms Control, 24–31 March 1986), p. 15; and John Pike, "Limitation on Space Weapons, A Preliminary Assessment" (Washington, D.C., Federation of American Scientists, February 1987, Mimeographed).
75. Lin, "New Weapon Technologies," note 64, pp. 84–86; William Durch, "Verification of Limitations on Antisatellite Weapons," in William C. Potter, ed., Verification and Arms Control (Lexington, Mass.: Lexington Books, 1985), p. 100.
76. For text, see Chapter 2, p. 69.
77. Compare with proposal by Sagdeev reported in Walter Pincus, "Soviet Says Talks Needed," Washington Post, 30 Oct. 1986, p. 4. (Discussed on p. 68, Chapter 2.)
78. Longstreth, et al., Impact, note 60, pp. 76–77.

4 ASSESSMENT AND CONCLUSIONS

The three policy paths discussed in the previous chapter—transition to a new offense/defense regime, force defense, and near-zero BMD under a strengthened ABM Treaty—yield four cases to compare: comprehensive area defense (competitive transition), light area defense (cooperative/virtual transition), force defense, and Treaty defense (folding together the accidental attack system and the virtual defense and near-zero cases). How do these cases compare in terms of their impact on the Treaty itself, on defense costs, on arms race stability, on crisis stability, on NATO cohension, and on U.S.-Soviet relations? Finally, what conclusions can be drawn and what recommendations made for U.S. policy?

IMPACT ON THE ABM TREATY

A competitive transition aimed at *comprehensive area defense* would destroy the Treaty and is more or less its antithesis, but a "broadly" interpreted agreement serves as a proxy for withdrawal, which could be useful to a competitive strategy to the extent that the Soviets agree. To date, they have been adamant in their disagreement.

The national *light area defense* of a virtual transition could be designed as a truly minimum system sufficient, in a zero-strategic-

141

ballistic-missiles world, only to defend against attacks mounted by a surreptitiously built offense, especially if an air-breathing offense remained the principal deterrent force. Equipped with ground-based, self-guided, ERIS-type interceptors, it might be designed around a relative handful of fixed sites. Alternatively, it might be designed as an expandable "cadre" system, with relatively few interceptors but with a full national radar base and warm launcher/interceptor production lines, or be built around rapidly deployable components, including radars. Such a cadre system would be expandable by definition. Although many of the current constraints in the Treaty might be maintained under this concept (e.g., limits on space-based components), any regulatory regime in force in a virtual transition would be very different in purpose from the current Treaty, sanctioning rather than precluding nationwide defense, and its expandability could prove destabilizing.

Force defense on a scale sufficient to defend single-silo ICBMs against a full-scale Soviet ICBM force would be so large as to effectively nullify the Treaty. Force defenses deployed to defend an offense reduced by one-half against a threat also reduced by one-half would not have appreciably less impact on the Treaty, unless the basing mode of the defended targets offered better leverage to the defense than do single-silo ICBMs. Defense of residual silo-based ICBMs might be able to take advantage of such leverage with fewer changes to the Treaty and, at least theoretically, with a boundable deployment in areas remote from major population centers.

Off-road mobile deployments of small ICBMs might benefit somewhat from a modest amount of BMD, given current threat levels, but any need for active defense would decline along with the threat, provided that mobile ICBM deployment areas remained large. BMD component production monitoring would be necessary to assure compliance with deployment ceilings. The more remote the deployment area(s) are from urban/industrial concentrations, and the more constrained their interceptors' range, the less stress on the Treaty regime.

The *Treaty defense* systems—zero to 100 interceptors in variations on the sorts of deployments currently allowed—would neither undercut the Treaty nor prevent efforts to shore it up, provided an accident-absorbing system could make do with the types and numbers of ABM components currently permitted. If, for example, a national ABM radar base were required to support the system, the risks it would entail for the Treaty regime and for breakout could outweigh the benefits of its deployment, as long as the probability of accidental attack is assessed to be low.

DEFENSE COSTS

This analysis has not independently assessed either the prospective total costs or the cost-effectiveness of the four BMD cases discussed above. How the cases would be ranked in terms of costs, however, seems reasonably clear. Comprehensive area defense would be by far the most costly. As Barry Blechman and Victor Utgoff observe in their study of national strategic defense systems, deploying and maintaining an extensive BMD (almost necessarily accompanied by air defense) would be equivalent in cost to creating and maintaining a fourth branch of the armed forces. A full-scale force defense, equivalent to the lowest layer or two of a comprehensive system, would cost perhaps one-fifth to one-tenth as much, in the range of $100 billion.[1] Blechman and Utgoff do not price a system like the light area defense discussed here, but relative to the two previous cases it need not be very costly; perhaps it would be equivalent in cost to that of the anti-accident system plus the cost of preparations to expand it, if necessary. The Treaty defense cases would each cost $1 or 2 billion annually to sustain a moderate level of BMD research, added to which would be the expenses of any offense survivability measures undertaken in lieu of BMD.

ARMS RACE IMPLICATIONS

Terminating the Treaty to pursue a competitive transition would reinforce MAD with ODD: the offense-defense dialectic. The U.S. and Soviet governments both have made it clear they they would offset any deployment by the other of BMD beyond Treaty limits with offense, at a minimum. Any policy that involved the development and deployment of an extensive strategic defense would also involve continuing system upgrades as the offensive and defense-suppressive threats change and evolve. An ODD world would at best be uncertain and expensive.

A virtual transition scenario seeks to avoid an arms race, and indeed to reverse the effects of the present one. However, it demands a good deal of trust that neither side will bolt the accord and seek to establish a strategic advantage. Phased-in defenses and rapid dismantlement of offense in a second stage of reductions would seek to minimize incentives to do so. But any significant disruption of the timing of this venture could lead to undesirable outcomes. Moreover, the responsive defense that would be the result of a successfully completed transition could respond equally well to the demands of a side's own breakout strategy.

Force defenses may or may not stimulate an offsetting offensive response, depending on the specific context of their deployment (e.g., agreed-upon versus unilateral deployment; or full-scale system versus capstone for cuts in offense). Deployed against an unconstrained offense, even force defense could be a recipe for an arms race. Deployed against constrained but still extensive offense, force defenses could leave the U.S. deterrent worse off than in the near-zero BMD case, all else being equal, unless the Soviet Union were willing to let U.S. defense deployments outstrip its own. If the objective of national policy is to maintain some number of survivable, deliverable ballistic missile RVs, then passive defense measures (in particular mobility on land and at sea) look much more attractive than BMD.

Near-zero BMD cannot guarantee that U.S.-Soviet arms competition will come to a dead halt, but at least it offers no incentives to build more offense.

CRISIS STABILITY

A competitive transition effort could have decidedly negative implications for crisis stability, as partially effective defenses were layered over large offensive forces that depended on the defense for their survivability (e.g., silo-based ICBMs). Tense U.S.-Soviet relations, expected under conditions of unrestrained military competition, could magnify the effect. Indeed, a deadly serious competition for advantage in defense could easily transform itself into competition for advantage per se, not only undermining crisis stability but helping to generate the crisis.

Light area defense in a virtual transition, especially in a situation that remained reliant for deterrence on air-breathing systems, should not decrease crisis stability, *provided* that cheating on offense and defense, that is, breakout potential, could be effectively minimized. But if such cheating could be reliably minimized, a transition-guarding defense should not be necessary.

Force defense is generally believed to enhance crisis stability, but if symmetric BMD deployments gave the Soviet Union the ability to absorb a high fraction of a U.S. retaliatory missile strike, then the impact of force defense on crisis stability would be at best ambiguous and at worst quite negative.

The near-zero/anti-accident case has neutral to positive implications for crisis stability—neutral to the extent that it neither helps nor hinders non-BMD measures to improve offense survivability, and positive to the extent that it inhibits deployment of a partially effective area defense, or absorbs an accidental attack that otherwise might spiral into war.

ALLIANCE RELATIONS

NATO and other U.S. allies have expressed interest in the possible technology windfalls that SDI might produce, while also expressing concern about SDI's possible end-product and its effect on U.S. willingness and ability to attend to its alliance commitments. Any factor that is seen as altering the basis either of extended deterrence, the sharing of risk, or the prospects for détente will put great stress on U.S.-allied relations. The Treaty is seen as supporting all three concepts and is strongly endorsed by all the major U.S. allies, in its traditional interpretation.[2] The alternative BMD cases can be judged as to their likely impact on NATO by their impact on the Treaty. France and Great Britain also have a direct military stake in the Soviet Union's continued observance of the Treaty. French and British missiles' pre-launch survivability would not benefit in any way from U.S. deployment of limited BMD, and their post-launch survivability would decrease in lockstep with Soviet missile defense effectiveness.

Strengthening the Treaty would be the path least likely to cause ripples in U.S.-NATO relations. The Treaty would then remain a symbol of ongoing efforts to moderate U.S.-Soviet competition to the net benefit of détente in Europe.

EAST-WEST RELATIONS

Pursuing a competitive transition strategy would set East-West relations back thirty years.[3] The probability that the Soviet defense establishment, political leadership, or economy would collapse in the face of a determined U.S. BMD challenge is low, as is the prospect that East-West cooperation would flourish in a determinedly ODD world.

At present, the Soviet Union also shows no interest in pursing a virtual transition strategy. It has agreed to the concept of offense reductions, but only on the condition that the Treaty be sustained for a decade and strengthened in its restrictions on space-based technology. Moreover, it has little interest in deploying BMD to defend ICBMs and even less interest in seeing the United States deploy such defenses, and already possesses an accident-absorbing ABM system. In short, reinforcing the strategic defense status quo is the only policy path in which the Soviet Union currently seems to have an interest. That interest is driven in part by Soviet perception of SDI, and restraint on SDI R&D is the Soviet Union's quid pro quo for bilateral reductions in offense. Until and unless the Reagan administration responds on that score, offense reductions, as well as other Treaty issues, are likely to remain in limbo.

Some believe that current Soviet interest in the ABM Treaty extends no further than one-sided restraint of U.S. programs while Soviet ABM programs move steadily forward, as often as not on the far side of that broad gray zone dividing fully compliant behavior from deliberate violations. Criticism of Soviet practice with regard to BMD and Treaty compliance is in part well taken. But to state that the Soviet Union is pursuing interests beneficial to itself under the ABM Treaty regime is to state the obvious. That it may cheat to further those interests if the odds are against detection is an unfortunate corollary, pointing up once again the need for effective intelligence, well-defined negotiating goals, and clear agreements.

THE 1987 TREATY REVIEW

Every five years, the ABM Treaty comes up for review by the two parties. This is the only timeline in the agreement, apart from the withdrawal clause. Although the Treaty may be amended by the parties at any time, the review periods provide a regular opportunity to focus government attention on the agreement. In 1977, at the first review conference, the two sides agreed "that the Treaty is operating effectively . . . serves the security interests of both parties, decreases the risk of outbreak of nuclear war, facilitates progress in the further limitation and reduction of strategic offensive arms, and requires no amendment at this time."[4]

In 1982, at the second review conference, the joint communiqué was much more perfunctory. Although each party "reaffirmed its commitment to the aims and objectives of the Treaty," and to the SCC consulting process, there was no joint assessment of the Treaty's effectiveness.[5]

The third review period opens in October 1987. Because of SDI and the Reagan administration's interest in BMD, the Treaty has been under more or less continuous U.S. review in recent years. As noted earlier, recent sessions of the Geneva Defense and Space Talks have included a working group on ABM Treaty definitional issues, but the U.S. delegation has been restricted to discussions consistent with the administration's broad interpretation of the Treaty. When Secretary of State Shultz met with Soviet Foreign Minister Shevardnadze in April 1987, the Soviet Union proposed more formal consideration of ABM Treaty issues, which could form part of the 1987 review, to take place in the fall of 1987 or early 1988.[6]

Substantive U.S. policy options for the 1987 review would range from a challenge of Soviet compliance practice as constituting material breach of the Treaty's provisions and grounds for U.S. withdrawal, to the non-committal approach used in 1982, to a major effort to settle the deployment and development issues discussed earlier. Procedural options include a decision on the venue for the review conference. Previous reviews took place in the SCC; the 1987 review could take place there or in the Defense and Space Talks. A certain amount of symbolism is embedded in this choice: the SCC is a forum for working out Treaty implementation and compliance issues; the Defense and Space Talks have been a forum for efforts to discuss a transition away from the current Treaty regime.

Withdrawal from the ABM Treaty would be a more complex and risky operation than cancelling informal observance of an unratified agreement such as SALT II: the ABM Treaty is in force and has more political support than SALT II. Withdrawal would require making a politically and militarily compelling case of Soviet material breach of the Treaty but, as such a case is as yet difficult to make, a withdrawal strategy could backlash badly.

The next option would be an effort to negotiate joint adoption of the *broad interpretation*. From a hardline policy perspective, the broad interpretation buys the best of two worlds: relaxed constraints on the highest BMD technologies, as well as continued constraints on testing and deployment of more traditional technologies, which would tend to favor U.S. development programs while keeping the most mature Soviet programs in check. The Soviet Union has shown no interest in accepting this package, but must consider its options carefully, since the United States, frustrated in its efforts to negotiate a broad interpretation, might be encouraged to move toward withdrawal. On the other hand, the political costs of withdrawal would suggest unilateral implementation of the broad interpretation as the more likely U.S. fallback, with the next move left up to Moscow. Since a Soviet move away from the Treaty would play to this hardline strategy, increasing the grounds for U.S. withdrawal, and sacrificing a good deal of Moscow's political leverage in Europe, the Soviets would do better to wrap themselves in the Treaty, leaving it to the allies to keep pressure on Washington, and the Congress to keep the SDI testing program consistent with the narrow interpretation. In that way, Gorbachev could retain the political high ground and maneuver the issue into a defeat for U.S. policy. To maintain that leverage, however, Gorbachev would have to appear willing to alleviate legitimate Western concerns about Soviet compliance with the Treaty.

Given the potential backlash of the first two options, a low-key review session with *non-committal* results along the lines of the 1982 communiqué would be an attractive alternative to hardliners with appeal to Treaty supporters as well, curing no ills but doing no damage. Should the Reagan administration be uncomfortable endorsing even the "aims and objectives" of the Treaty as it did in 1982, the review conference communiqué need only acknowledge that the required review took place.

Finally, a *problem-solving* review conference would be the sort that ABM Treaty supporters might prefer and that Moscow seems to be advocating. A problem-solving conference could aim at a number of outcomes: it could seek joint, public endorsement of the traditional interpretation of the Treaty and pledges of non-withdrawal for a five- to ten-year period. It could further seek a resolution of the

LPAR deployment issue; agreed definitions of Treaty-relevant development and testing, including definitions of threshold ABM capabilities; and quantitative, operational definitions of testing in an ABM mode.

Given the orientation of the Reagan administration toward the Treaty and SDI, however, such a problem-solving approach on its part is very unlikely. Nor is it clear that the Soviet Union would, in fact, be willing to commit itself to new testing thresholds, although in meetings with non-governmental delegations of U.S. arms control experts, Soviet interlocutors have suggested as much. The suggestions have yet to be tested.

The best to be hoped for from the 1987 review conference from the Treaty's perspective, then, may be a quiet, non-committal outcome. At this stage of the debate over the ABM Treaty's future, it is likely that the more visible the review, the less good it would do the agreement.

CONCLUSIONS

The ABM Treaty today is under siege of ideology, technology, and time.

The ideological attack portrays the Treaty as a vestige of a lost and unlamented era of American weakness and unwillingness to meet the Soviet challenge, and as the guardian of the "dogma" of mutual vulnerability. If Western strategic doctrine is in a box, then the Treaty, from this perspective, is the lid.

The technological assault is both evolutionary and revolutionary. In the former category are incremental changes to familiar technologies that promise to make them increasingly hard to distinguish from BMD components. The "revolutionary" category includes development of lasers, particle beams, electromagnetic railguns, microwave generators, and doubtless other programs hidden in the blanked-out lines of Congressional authorizations.

Time, of course, is the enemy of everything. Arms control agreements, being political barriers to defense business-as-usual, are subject to erosion in proportion to the real constraints they impose on weapons whose development, testing, and deployment is seen by some to be feasible and, even, desirable.

There is as yet no consensus that a better security payoff can be gleaned by either party by withdrawing from the Treaty, but Moscow presses the limits of the agreement while appeals to a better, unilateral security payoff through SDI are being made in the United States, not only by outside critics of arms control but also by the Reagan administration itself. Since at least 1983, the effect of Reagan administration actions and rhetoric has been to set the United States upon the path to Treaty termination, a path not in U.S. interests.

Politics and ideology drive U.S. and Soviet strategic nuclear arms, not the reverse. So long as one side views the other as its primary security threat, neither will allow the other the luxury of unchallenged security, and efforts to build a security bubble are likely to be frustrated. On the other hand, if prevailing views change, security might be had without a bubble, at a fraction of the current cost.

If we feel compelled to deploy BMD, let us consider some insurance against accidental missile attack—technically feasible in a few years with non-nuclear interceptors, and deployable without further negotiation if we make use of the Grand Forks deployment area and current radar nets suffice to vector long range interceptors. Such a system would be hard to justify on cost-effectiveness grounds due to the low probability of accidental missile attack. Nonetheless, it might be useful for the political symmetry it offers with the Moscow system, and for the hands-on experience with BMD systems operation that it would provide and that Treaty critics can now claim is lacking. Not a military necessity, it should be considered for deployment only within current Treaty constraints.

The ABM Treaty has served U.S. and Western security interests well and it will continue to do so. It can and should be maintained and made ready for the twenty-first century.

The historian John Lewis Gaddis, in his careful assessment of the "long peace" that has prevailed among the world's great powers since 1945, argues that the nuclear weapon, which poses great risks to all states, is the single most important factor that sets this era apart from all other modern interwar periods. The existence of nuclear weapons robs statesmen of the optimism they need to launch major wars.[7] Aggressors plan to win; against nuclear retaliation they cannot win. That is an essential truth of the nuclear age. Before we permit the rekindling of that lost sense of optimism with systems that provide, at great cost, an illusion of survivability, we should think very carefully about the possible consequences.

NOTES

1. See Blechman and Utgoff, note 17 (Chapter 2) for estimated costs of air and missile defenses. By their estimate, ten-year costs for extensive defenses hover at the $500 to $800 billion range (fiscal 1987 dollars). Costs of force defenses range from roughly $75 to $160 billion, depending on whether air defense is included. For a summary of the study, see Barry M. Blechman and Victor A. Utgoff, "The Macroeconomics of Strategic Defense," *International Security* (Winter 1986-87): 33-70.

2. See Ivo Daalder, *The SDI Challenge to Europe* (Cambridge, Mass.: Ballinger Publishing Company, 1987).

3. This is not a figure of speech. Those who pine for the 1950s have not examined that period in detail or viewed it through the perceptual lenses of the time. See Richard K. Betts, "A Nuclear Golden Age? The Balance Before Parity," *International Security* (Winter 1986-87): 3-32.

4. U.S. Arms Control and Disarmament Agency, *Documents on Disarmament, 1977.* Communiqué of the U.S.-Soviet Standing Consultative Commission, 21 Nov. 1977, p. 791.

5. U.S. Department of State, "SCC Completes Review of ABM Treaty," *Daily Bulletin*, 16 Dec. 1982, p. 11.

6. R. Jeffrey Smith, "USSR Said to Request Parley on ABM Limits," *Washington Post*, 26 Apr. 1987, p. 1.

7. John Lewis Gaddis, "The Long Peace: Elements of Stability in the Postwar International System," *International Security* (Spring 1986): 121.

INDEX

153

ABOUT THE AUTHOR

William J. Durch is assistant director of the Defense and Arms Control Studies Program, Massachusetts Institute of Technology, coordinator of the MIT/Harvard Summer Program on Nuclear Weapons and Arms Control, and a doctoral candidate in political science (defense studies) at MIT. From 1981 to 1983 he was research fellow and project coordinator at the Center for Science and International Affairs, Harvard University, where he chaired the Space Policy Working Group. From 1978 to 1981 he served in the U.S. Arms Control and Disarmament Agency (ACDA), working on naval air defense issues, conventional arms transfer, control of anti-satellite weapons, and ballistic missile defense. From 1973 to 1978 he was on the staff of the Center for Naval Analyses. He holds an M.A. in international politics from the George Washington University and a B.S.F.S. from the Georgetown University School of Foreign Service.